A GENERIC FAULT-TOLERANT ARCHITECTURE
FOR REAL-TIME DEPENDABLE SYSTEMS

A Generic Fault-Tolerant Architecture for Real-Time Dependable Systems

Edited by

David Powell

CNRS, France

KLUWER ACADEMIC PUBLISHERS

BOSTON / DORDRECHT / LONDON

A C.I.P. Catalogue record for this book is available from the Library of Congress.

ISBN 978-1-4419-4880-9

Published by Kluwer Academic Publishers,
P.O. Box 17, 3300 AA Dordrecht, The Netherlands.

Sold and distributed in North, Central and South America
by Kluwer Academic Publishers,
101 Philip Drive, Norwell, MA 02061, U.S.A.

In all other countries, sold and distributed
by Kluwer Academic Publishers,
P.O. Box 322, 3300 AH Dordrecht, The Netherlands.

Printed on acid-free paper

Table of Contents

List of Figures

List of Tables

List of Contributors

Jean Arlat .... LAAS-CNRS
Arturo Amendola .... Ansaldo Segnalamento Ferroviario
Berthold Attermeyer .... Siemens AG Österreich PSE
Cinzia Bernadeschi .... PDCC-Università di Pisa
Ljerka Beus-Dukic[1] .... University of York
Jean-Paul Blanquart .... Matra Marconi Space France
Andrea Bondavalli[2] .... PDCC-CNUCE-CNR
Silvano Chiaradonna .... PDCC-CNUCE-CNR
Paolo Coppola .... Intecs Sistemi
Carlo Dambra .... Ansaldo Segnalamento Ferroviario
Yves Deswarte .... LAAS-CNRS
Felicita Di Giandomenico .... PDCC-IEI-CNR
Alessandro Fantechi[2] .... PDCC-IEI-CNR
Stefania Gnesi .... PDCC-IEI-CNR
Fabrizio Grandoni .... PDCC-IEI-CNR
Eric Jenn .... Technicatome
Karama Kanoun .... LAAS-CNRS
Stéphane Lautier .... Technicatome
Ivan Mura[3] .... PDCC-IEI-CNR
Vincent Nicomette[4] .... Matra Marconi Space France
Alessandro Paganone .... Intecs Sistemi
David Powell .... LAAS-CNRS
Christophe Rabéjac .... Matra Marconi Space France
Mohamed Tahar Jarboui .... LAAS-CNRS
Vincent Thevenot .... Technicatome
Eric Totel .... Matra Marconi Space France
Andy Wellings .... University of York

---

[1] Now with University of Northumbria at Newcastle
[2] Now with Università di Firenze
[3] Now with Motorola Technology Center Italy
[4] Now with LAAS-CNRS

Foreword

The design of computer systems to be embedded in critical real-time applications is a complex task. Such systems must not only guarantee to meet hard real-time deadlines imposed by their physical environment, they must guarantee to do so dependably, despite both physical faults (in hardware) and design faults (in hardware or software). A fault-tolerance approach is mandatory for these guarantees to be commensurate with the safety and reliability requirements of many life- and mission-critical applications.

This book explains the motivations and the results of a collaborative project[1], whose objective was to significantly decrease the lifecycle costs of such fault-tolerant systems. The end-user companies participating in this project already deploy fault-tolerant systems in critical railway, space and nuclear-propulsion applications. However, these are proprietary systems whose architectures have been tailored to meet domain-specific requirements. This has led to very costly, inflexible, and often hardware-intensive solutions that, by the time they are developed, validated and certified for use in the field, can already be out-of-date in terms of their underlying hardware and software technology.

The project thus designed a *generic* fault-tolerant architecture with two dimensions of redundancy and a third multi-level integrity dimension for accommodating software components of different levels of criticality. The architecture is largely based on commercial off-the-shelf (COTS) components and follows a software-implemented approach so as to minimise the need for special hardware. Using an associated development and validation environment, system developers may configure and validate *instances* of the architecture that can be shown to meet the very diverse requirements of railway, space, nuclear-propulsion and other critical real-time applications. This book describes the rationale of the generic architecture, the design and validation of its communication, scheduling and

---

[1] Esprit project n°20716: GUARDS: a Generic Upgradable Architecture for Real-time Dependable Systems.

fault-tolerance components, and the tools that make up its design and validation environment.

The book is organised as follows.

Chapter 1 gives a general introduction to the project and a general overview of the proposed generic fault-tolerant architecture, together with its associated development and validation environments.

Chapters 2 to 6 detail the architecture. Chapter 2 is dedicated to the inter-channel communication network, which constitutes the central articulation point of the architecture. Chapter 3 describes the approach that has been followed for scheduling replicated real-time task executions and inter-channel communication. Chapter 4 details the adopted fault-tolerance approach, describing in particular the diagnosis and state recovery mechanisms. Chapter 5 addresses the problem of combining the redundant outputs of the fault-tolerant architecture into a single effect on the external world. Chapter 6 describes the architecture's innovative multilevel integrity mechanisms, which allow the architecture to support application software of different levels of criticality.

Chapters 7 to 9 are devoted to the tools and activities that are necessary for the development and validation of the architecture. Chapter 7 describes the architecture development environment, which consists of a set of tools for designing instances of the generic architecture. Chapter 8 outlines the formal verification activities that were carried out on various aspects of the architecture. Chapter 9 then reports the various modelling activities aimed at characterising and assessing the achieved dependability.

Chapter 10 concludes the book with a description of three prototype systems that have been developed to demonstrate the proposed approach.

Chapter 1

Introduction and Overview

The development and validation of fault-tolerant systems for critical real-time applications are currently both costly and time-consuming. Most such systems developed until now have been specialised to meet the particular requirements of the application domain for which they were targeted. This specialisation has led to very costly, inflexible, and often hardware-intensive solutions that, by the time they are developed, validated and certified for use in the field, can already be out-of-date in terms of their underlying hardware and software technology. This problem is exacerbated in application domains that require the systems to be deployed for several decades, i.e., almost an order of magnitude longer than the typical lifetime of a generation of computing technology. Furthermore, it is currently very difficult to re-use the results of previous fault-tolerance developments when building new products.

To tackle these issues, a consortium of European companies and academic partners was formed (see annex A) to design and develop a Generic Upgradable Architecture for Real-time Dependable Systems (GUARDS, ESPRIT project n°20716), together with an associated development and validation environment. This book describes the project's results.

The end-user companies in the project consortium all deploy ultra-dependable real-time embedded computers in their systems, but with very different requirements and constraints resulting from the diversity of their application domains: nuclear propulsion, railway and space systems. The considered applications are *hard real-time* applications in that the respect of deadlines is a condition for mission success.

The architecture developed by the project aims to significantly decrease the lifecycle costs of such embedded systems. The intent is to be able to configure *instances* of a generic architecture that can be shown to meet the very diverse requirements of these (and other) critical real-time application domains. Moreover, to take advantage of rapid technological progress, another fundamental motivation of the generic architecture is to facilitate system upgrades through the use, whenever possible, of commercial off-the-shelf (COTS) components.

D. Powell (ed.), A Generic Fault-Tolerant Architecture for Real-Time Dependable Systems, 1–26.

Validation and certification of safety-critical and safety-related systems represent a high proportion of their overall life-cycle costs. The proposed generic architecture addresses these costs by following a "design for validation" approach, focussing validation obligations on a minimum set of critical components. In line with this approach, the architecture offers support for software components of different levels of criticality by enforcing an integrity policy. This ensures that the operation of thoroughly validated (i.e., high integrity) critical components is not endangered by residual design faults in less well-validated components of lower criticality, e.g., non-certified COTS components. Finally, the generic aspect of the architecture allows already-validated components to be re-used in different instances.

The remainder of this chapter is structured as follows. The first two sections outline the requirements of the architecture in terms of the considered application domains and the proposed design rationale. The architecture itself is defined in Section 1.3. Central to the architecture is an inter-channel communication network, which is described in Section 1.4. Section 1.5 details the inter-channel fault-tolerance mechanisms while Section 1.6 discusses the scheduling issues raised by active replication of real-time tasks. Sections 1.8 and 1.9 discuss respectively the development and validation environments that accompany the architecture. Finally, Section 1.10 describes some domain-specific instances of the architecture.

1.1 Application Domains

The architecture is aimed at three major classes of applications: mission-critical embedded systems, safety-critical and safety-related process control systems, and critical instrumentation and information display systems. The architecture can potentially address a wide variety of domains, including avionics, defence systems, chemical, petroleum and processing industries, etc., but the primary focus is that of the application domains of the project consortium's end-user partners, which are diverse and rich enough to validate the proposed approach:

- Railway applications.
- Nuclear applications.
- Space applications.

We describe below some typical railway, nuclear and space applications, along with their main characteristics.

1.1.1 Railway Applications

In the railway domain, the GUARDS architecture can be applied to at least five different functions and products:

- Communication technologies, especially for track to train communication.
- Information processing technologies.
- Signalling system component technologies.

- Train control products.
- Process technologies used in developing signalling and train control products.

A typical instance of the architecture in the railway domain would be a fail-safe control system. Railway standards dictate extremely low catastrophic failure rates for individual subsystems (e.g., less than 10^{-11}/hour with respect to physical faults). Furthermore, in railway applications, it is common to physically segregate subsystems responsible for vital (safety-critical) functions from those responsible for non-vital functions. However, the project decided to investigate the possibility of a single instance supporting both high-integrity vital functions and low-integrity non-vital functions since this would offer significant economic advantages.

1.1.2 Nuclear Applications

In the nuclear domain, GUARDS targets mainly the Instrumentation and Control (I&C) system for submarine propulsion, and more specifically the subsystem related to process regulation, to control and monitoring of actuators that are critical for the availability of power generation, and the man-machine interfaces in the control room. Generally speaking, this covers the equipment that is classified B according to [IEC 61226], i.e., equipment that is critical for the operational mission. Other possible targets include:

- Experimental reactors.
- Nuclear I&C revamping (i.e., a major upgrade in the product life).

These kinds of application typically impose very long mission durations between possible maintenance operations.

Two constraints from this application domain impose quite severe restrictions on the design space. First, it must be possible to separate redundant elements of the architecture by several tens of meters so as to tolerate physical damage. Second, to avoid obsolescence during the submarine's lifetime, the use of unmodified COTS operating systems is mandatory.

1.1.3 Space Applications

In the space domain, there are many critical real-time functions embedded in systems aimed at long life-time missions or manned transportation. These functions must be supported by a highly-dependable infrastructure. Examples of such applications are:

- Autonomous spacecraft and long life-time scientific probes (e.g., the Rosetta probe, which is planned to be launched in 2003 and rendezvous with the comet Wirtanen in 2012).
- Space systems tolerant to very harsh environments (military systems or robots).
- Automatic systems performing rendezvous with manned systems (e.g., the European Space Agency's planned Automated Transfer Vehicle (ATV)).

- Manned space vehicles and platforms (e.g., a Crew Transport or Return Vehicle (CTV/CRV)).

A particularly challenging application in the space domain is that of an autonomous spacecraft carrying out missions containing phases that are so critical that tolerance of several faults may be required (e.g., target fly-by or docking). During non-critical phases, the redundant elements may be powered down to save energy. Moreover, as already considered for railway applications, it is necessary for an instance to be able to support software of different integrity levels: high-integrity critical software that is essential for long-term mission reliability and potentially unreliable payload software.

1.2 Design Rationale

In this section, we first discuss the genericity of the proposed approach and the constraints imposed by the development process. We then discuss the requirements of the architecture with respect to tolerable fault classes and real-time scheduling.

1.2.1 Genericity

To merit the epithet "generic", the architecture should be configurable to meet the widest possible spectrum of dependability and real-time requirements. The idea is to define a set of mechanisms, implemented as GUARDS hardware and software components, among which it is possible to pick and choose those that are necessary to implement a particular instance. Ideally, it should be possible to implement and validate the functional, application-specific components and the GUARDS components separately.

Instances of the architecture may adopt different COTS real-time operating systems and COTS hardware. To maximise flexibility, it is therefore necessary for the GUARDS components to be as independent as possible from the underlying infrastructure. To this end, the GUARDS software components should have well-defined interfaces based on a common set of services proposed by the supported real-time operating systems. The cost of adaptation to another operating system should be as limited as possible. We therefore decided to define the common set of services according to the POSIX standard [IEEE 1003]. Choosing such a standard does not guarantee the architecture to be future-proof, but it does ensure a certain level of reproducibility and the capability for the compliance of new components to be checked. If POSIX becomes obsolete sooner than expected, an emulation of the services should still be possible on the basis of a new standard.

1.2.2 Development Process

A development and validation environment should accompany the architecture to assist system designers in configuring instances to suit their requirements. However, different development methods, languages and tools are used at the application level. These methods and tools are tightly linked to the considered application domain (e.g., the specification methods are different in the space, railway and nuclear domains). Therefore, although it is not possible to define a common set of tools for developing and validating instances of the architecture, the various tools should ideally be able to communicate with each other and together constitute a consistent engineering environment. Furthermore, it should be possible to interface the GUARDS software components with application modules written in different languages.

To be able to assess the dependability of instances of the architecture, the validation environment should include tools able to support a set of pre-defined models and the means to specify new models adapted to the application. Experimental validation of prototype instances by fault injection also requires specific tools supporting the whole fault injection process: injection point specification, injection scenarios, result logging and storage, etc.

1.2.3 Fault Classes

The architecture should be able to tolerate permanent and temporary physical faults (of both internal and external origins) and should provide tolerance or confinement of software design faults. This wide spectrum of fault classes [Laprie *et al.* 1995] has several consequences beyond the basic physical redundancy necessary to tolerate permanent internal physical faults. Tolerance of permanent external physical faults (e.g., physical damage) requires geographical separation of redundancy. Temporary external physical faults (transients) can lead to rapid redundancy attrition unless their effects can be undone. This means that it should be possible to recover corrupted processors. Temporary internal physical faults (intermittents) need to be treated as either permanent or transient faults according to their rate of recurrence.

Many design faults can also be tolerated like intermittents if their activation conditions are sufficiently decorrelated [Gray 1986] (e.g., through loosely-coupled replicated computations). However, design faults that are activated systematically for a given sequence of application inputs can only be tolerated through diversification of design or specification. Due to limited resources, the project has not considered diversification of application software beyond imposing the requirement that no design decision should preclude that option in the future. However, we have studied the use of integrity level and control-flow monitoring mechanisms to ensure that design faults in non-critical application software do not affect critical applications. Moreover, we have considered diversification for tolerating design faults in COTS operating systems. We have also encouraged activation condition decorrelation so as to provide some tolerance of design faults in replicated hardware and replicated applications.

1.2.4 Real-time

Since the considered applications impose hard real-time constraints, the architecture should be designed such that it is possible to ensure, by analysis, that the real-time behaviour of the application software is timely. In keeping with the genericity objective, the architecture should be capable of supporting a range of real-time computational and scheduling models that allow this objective to be reached.

The *computational model* defines the form of concurrency (e.g., tasks, threads, asynchronous communication, etc.) and any restriction that must be placed on application programs to facilitate their timing analysis (e.g., bounded recursion). Applications supported by GUARDS may conform to a time-triggered, event-triggered or mixed computational model.

The *scheduling model* defines the algorithm for allocating and ordering the system resources, with an associated timing analysis method. Three such models have been considered [Wellings *et al.* 1998]:

- Cyclic — as typified by the traditional cyclic executive.
- Cooperative — where an application-defined scheduler and the prioritised application tasks explicitly pass control between themselves to perform the required dispatching.
- Pre-emptive — the standard pre-emptive priority scheme.

We have focused primarily on the pre-emptive scheduling model since this is the most flexible and the one that presents the greatest challenges.

1.3 The Generic Architecture

The diversity of end-user requirements and envisaged fault-tolerance strategies led us to define a generic architecture that can be configured into a wide variety of instances. The architecture favours the use of COTS hardware and software components, with application-transparent fault-tolerance implemented primarily by software. Drawing on experience from systems such as SIFT [Melliar-Smith & Schwartz 1982], MAFT [Kieckhafer *et al.* 1988], FTPP [Harper & Lala 1990] and Delta-4 [Powell 1994], the generic architecture is defined along three dimensions of fault containment (Figure 1.1) [Powell 1997]:

- Channels, or *primary* physical-fault containment regions.
- Lanes, or *secondary* physical-fault containment regions.
- Integrity levels, or design-fault containment regions.

A particular instance of the architecture is defined by the dimensional parameters $\{C, L, I\}$ and a fault-tolerance strategy, leading to an appropriate selection of generic hardware and software GUARDS components. These generic components implement mechanisms for:

- Inter-channel communication.
- Output data consolidation.
- Fault-tolerance and integrity management.

Fault-tolerance and integrity management are software-implemented through a distributed set of generic system components (shown as a "middleware" layer on Figure 1.1). This layer is itself fault-tolerant (through replication and distribution of its components) with respect to faults that affect channels independently (e.g., physical faults). However, the tolerance of design faults in this system layer has not been explicitly addressed[1].

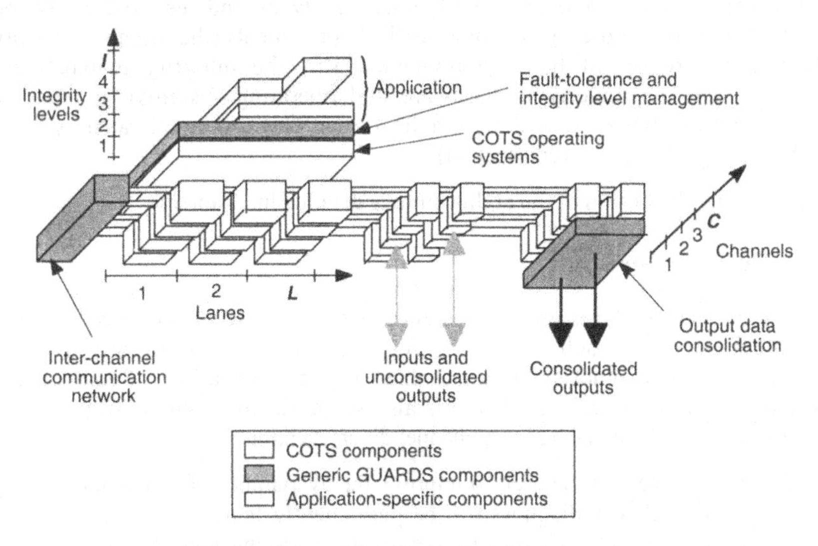

Figure 1.1 — The Generic Architecture

1.3.1 The Integrity Dimension

The integrity dimension aims to provide containment regions with respect to software design faults. The intent is to protect critical components from the propagation of errors due to residual design faults in less-critical components. Each application object is classified within a particular integrity level according to how much it can be trusted (the more trustworthy an object is, the higher its integrity level). The degree to which an object can be trusted depends on the available evidence in support of its correctness. In particular, the more critical an object (i.e., the more severe the consequences of its failure), the higher must be its integrity.

The required protection is achieved by enforcing an integrity policy to mediate the communication between objects of different levels. Basically, the integrity policy seeks to prohibit flows of information from low to high integrity levels, like in the Biba policy [Biba 1977]. However, this approach is inflexible. An object can obtain data of higher integrity than itself, but the data must then inherit the level of integrity of this object. This results in a decrease in the integrity of the data, without any possibility of restoring it. We deal with this drawback by providing special objects (*Validation Objects*) whose role is to apply fault-tolerance mechanisms on information flows. The purpose of these objects is to output reliable information by using possibly corrupted data as input (i.e., with a low integrity level). Such objects upgrade the trustworthiness of data and hence allow information flows from low to high integrity levels [Totel *et al.* 1998].

---

[1] Note, however, that correlated faults have been included in the models used to assess the dependability of instances of the architecture: see Chapter 9.

It must be ensured that it is not possible to by-pass the controls put into place to enforce the policy. This is achieved by spatial and temporal isolation, which can be provided respectively by memory management hardware and resource utilisation budget timers. Furthermore, for the most critical components (the topmost integrity level) and a core set of basic components (i.e., the integrity management components and the underlying hardware and operating systems), it must be assumed either that there are no design faults, or that they can be tolerated by some other means (e.g., through diversification).

The integrity dimension of the architecture is detailed in Chapter 6.

1.3.2 The Lane Dimension

Multiple processors or *lanes* are used essentially to define secondary physical fault containment regions. Such secondary regions can be used to improve the capabilities for fault diagnosis within a channel, e.g., by comparison of computation replicated on several processors. There is also scope for improving coverage with respect to design faults by using intra-channel diversification.

Alternatively, lanes can be used to improve the availability of a channel, e.g., by passivating a node that is diagnosed to be permanently faulty. The required fault diagnosis could be triggered either by the error-processing mechanisms within a channel or through an error signal from the inter-channel voting mechanisms.

Further reasons for defining an instance with multiple lanes include parallel processing to improve performance and isolation of software of different integrity levels. To aid the timing analysis of such software we require that the multiple processors within a channel have access to shared memory (see Section 1.7).

1.3.3 The Channel Dimension

Channels provide the *primary* fault containment regions that are the ultimate line of defence within a single instance for physical faults that affect a single channel. Fault tolerance is based on active replication of application tasks over the set of channels. It must be ensured that replicas are supplied with the same inputs in the same order, despite the occurrence of faults. Then, as long as replicas on fault-free channels behave deterministically, they should produce the same outputs. Error processing can thus be based on comparison or voting of replica outputs.

Not all instances require the same number of channels. In fact, one could imagine an instance with just one channel. This would be the case for an application that only requires multiple integrity levels, or for which the fault-tolerance mechanisms implemented within a channel are judged to be sufficient. It should be expected, however, that most applications require instances with several channels. Important cases are:

- Two channels: motivated either by a requirement for improved safety (using inter-channel comparison) or improved reliability (based on intra-channel self-checking to provide crash failure semantics).

- Three channels: the well-known triple modular redundancy (TMR) strategy that enables most[2] faults in one channel to be masked. In addition, any disagreements are detected and used as inputs for error diagnosis and fault treatment.
- Four channels: to enable masking of completely arbitrary faults[3] or to allow a channel to be isolated for off-line testing while still guaranteeing TMR operation with the remaining on-line channels.

Instances of the architecture with more than four channels have not been envisaged.

1.4 Inter-Channel Communication Network

Central to the architecture is an inter-channel communication network (ICN), which fulfils two essential functions:

- It provides a global clock to all channels.
- It allows channels to achieve interactive consistency (consensus) on non-replicated data.

The ICN consists of an ICN-manager for each channel and unidirectional serial links to interconnect the ICN-managers. An ICN-manager can simultaneously broadcast data to the remote ICN-managers over an outgoing serial link and receive data from the remote ICN-managers over the other links.

1.4.1 Clock Synchronisation

The ICN-managers constitute a set of fully interconnected nodes. Each node has a physical clock and computes a global logical clock time through a fault-tolerant synchronisation algorithm. Since COTS-based solutions are preferred within GUARDS, we focused on software-implemented algorithms. In particular, we considered both convergence averaging and convergence non-averaging algorithms [Ramanathan *et al.* 1990].

The GUARDS architecture uses a convergence-averaging solution based on [Lundelius-Welch & Lynch 1988] and applied to up to four nodes (i.e., ICN-managers in our architecture). This choice was motivated mainly by reasons of performance and design simplicity. However, this class of algorithm cannot tolerate Byzantine-faulty clocks in a three-channel configuration. Therefore, the probability of occurrence of a Byzantine clock should be carefully considered in this case. This

---

[2] The exception is that of Byzantine clock behaviour (see Section 1.4.1).

[3] By "arbitrary" we mean that the *communication* behaviour of a faulty channel is in no way constrained: it may omit to send messages, delay them, or even send conflicting messages to the other channels (i.e., the so-called "Byzantine behaviour" [Lamport *et al.* 1982]). Strictly speaking, however, there must be some restrictive assumption about what a faulty channel can do: for instance, it must be assumed that a faulty channel cannot affect the system *structure* in any way, e.g., by destroying another channel [Cachin *et al.* 2000].

probability is expected to be negligible, since the ICN serial links are broadcast media and the ICN-managers can check that they receive a syntactically correct synchronisation message in a well-defined local time window.

The global clock maintained by the set of ICN-managers is broadcast, via the intra-channel back-plane busses, to the processors and I/O boards local to a channel.

1.4.2 Interactive Consistency

The issue of exchanging private data between channels and agreeing on a common value in the presence of arbitrary faults is known as the interactive consistency problem (the symmetric form of the Byzantine agreement problem) [Pease *et al.* 1980]. The two fundamental properties that a communication algorithm must fulfil to ensure interactive consistency are:

- *Agreement*: if channels p and q are non-faulty, then they agree on the value ascribed to any other channel.
- *Validity*: if channels p and q are non-faulty, then the value ascribed to p by q is indeed p's private value.

In the general case, the necessary conditions to achieve interactive consistency in spite of up to m arbitrarily faulty channels are [Lala & Harper 1994]:

- At least $3m+1$ channels.
- At least $2m+1$ disjoint inter-channel communication links.
- At least $m+1$ rounds of message exchange.
- Bounded skew between non-faulty channels.

Under the assumption of authenticated messages, which can be copied and forwarded but not undetectably altered by a relayer, the condition on the minimal number of channels can be relaxed to $m+2$. Nevertheless, at least $2m+1$ channels are still necessary if majority voting must be carried out between replicated application tasks.

The interactive consistency protocol used in GUARDS is based on the ZA algorithm [Gong *et al.* 1998], which was derived from the Z algorithm [Thambidurai & Park 1988] by adding the assumption of authentication. In particular, authentication precludes the design fault in the Z algorithm identified in [Lincoln & Rushby 1993]. Following the hybrid fault model described in [Lincoln & Rushby 1993], the protocol allows for both arbitrarily faulty channels and channels affected by less severe kinds of faults (e.g., omission faults).

For performance reasons, and since by assumption the architecture only needs to tolerate accidental faults and not malicious attacks, we preferred to use a keyed checksum scheme for message authentication rather than resorting to true cryptographic signatures. Under this scheme, multiple checksums are appended to each (broadcast) message. Each checksum is computed over the concatenation of the data part of the message and a private key that is known only to the sender and to one of the broadcast destinations.

1.4.3 Scheduling

The ICN is scheduled according to a table-driven protocol. The schedule consists of a *frame* (corresponding to a given application mode) that is subdivided into *cycles* and *slots*. The last slot of a cycle is used for clock synchronisation, so the length of a cycle is fixed either by the required channel synchronisation accuracy or by the maximum I/O frequency in a given mode. The other slots of a cycle are of fixed duration and can support one fixed-sized message transmission (and up to three message receptions). In the current implementation, each message may contain 1000 bytes.

Further details on the inter-channel communication network and its implementation are given in Chapter 2. Scheduling of the ICN is detailed in Chapter 3 and in Chapter 7.

1.5 Inter-Channel Error Processing and Fault Treatment

From a conceptual viewpoint, it is common to consider fault tolerance as being achieved by error processing and fault treatment [Anderson & Lee 1981]. *Error processing* is aimed at removing errors from the computational state, if possible before failure occurrence. In general, error processing involves three primitives [Laprie 1992]: error detection, error diagnosis and error recovery. *Fault treatment* is aimed at preventing faults from being activated again and also involves three primitives [Laprie 1992]: fault diagnosis, fault passivation and reconfiguration.

In GUARDS, error recovery is achieved primarily by error compensation, whereby the erroneous state contains enough redundancy to enable its transformation into an error-free state. This redundancy is provided by active replication of critical applications (although diversification is not precluded) over the C channels. Replication is managed by software, including comparison or voting of computed results, and is transparent to the application programmer. Error processing thus relies primarily on N-modular redundancy to detect disagreeing channels and (when $C \geq 3$) to mask errors occurring in the voted results at run-time. When $C=2$, two possibilities are offered, as already mentioned in Section 1.3.3:

- Error detection (locally by a channel) and compensation (by switching to a single channel configuration).
- Error detection (by channel comparison) and switching to a safe state (a degenerate form of forward recovery).

Figure 1.2 illustrates the replicated execution of an iterative task in the case of a three-channel configuration. After reading the replicated sensors, the input values are consolidated across all channels after a two-round interactive consistency exchange over the ICN. The application tasks are then executed asynchronously, with pre-emptive priority scheduling allowing different interleavings of their executions on each channel. This diversifies the activities of the different channels, thereby allowing many residual design faults to be tolerated as if they were intermittents (cf. Section 1.2.3). It also facilitates the execution of non-replicated tasks.

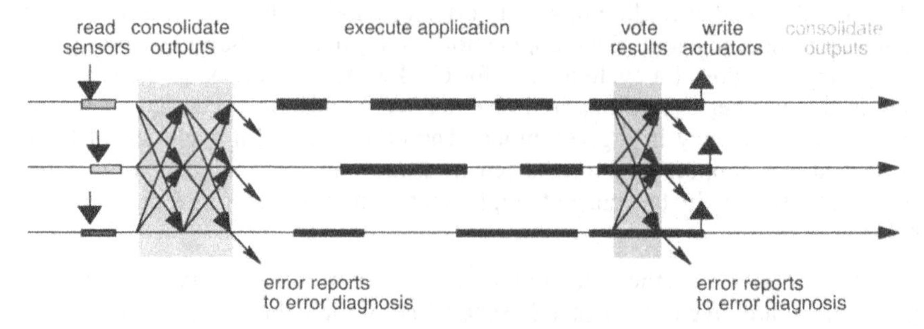

Figure 1.2 — TMR Execution of an Application Function split into Sequential Threads

Application *state variables* (which contain values that are carried over between iterations) are used together with consolidated inputs to compute the output values, which are exchanged in a single round over the ICN and voted. The voted results are then written to the actuators, possibly via output consolidation hardware (cf. Figure 1.1 and Section 1.6), which allows the physical values to be voted.

Since neither the internal state variables of the underlying COTS operating systems nor the totality of the application state variables are voted, further error recovery is necessary to correct any such state that becomes erroneous (note that this may be the case even in the event of a transient fault). However, this is a secondary, non-urgent error recovery activity since, until another channel is affected by a fault, the error compensation provided by output voting or switching can be relied upon to ensure that correct outputs are delivered to the controlled process. Consequently, this secondary error recovery can be viewed as part of fault treatment.

In the next section, we describe the GUARDS diagnosis mechanisms, which include both error diagnosis, to decide whether the damage to a channel's state warrants further action, and fault diagnosis to decide the location and type of the fault and thus the necessary corrective action. Then, in Section 1.5.2, we describe the state recovery procedure that allows re-integration of a channel after a transient fault or repair of a permanent fault.

1.5.1 Diagnosis

The first step in diagnosis is to collect error reports generated during the interactive consistency and consolidation exchanges (majority voting discrepancies, timing errors, ICN bus transmission errors, protocol violations, etc.) and then to filter them to assess whether the extent of damage warrants further action. Indeed, some reported errors may not have resulted in any change to the state of a channel. Alternatively, if only a small part of the state has become erroneous, then an erroneous channel might correct itself autonomously by overwriting the erroneous variables during continued execution. If such fortuitous recovery does not occur, an explicit forward recovery action is necessary to reconstruct a correct state.

The filtering of errors is done using a software-implemented mechanism known as an α-count, which was originally proposed for the discrimination of transient versus intermittent-permanent faults [Bondavalli *et al.* 1997a]. Error reports are processed on a periodic basis, giving lower weights to error reports as they get older. A score variable α_i is associated to each not-yet-removed component i to record information about the errors experienced by that component. α_i is initially set to 0, and accounts for the L-th judgement as follows:

$$\alpha_i(L) = \begin{cases} \alpha_i(L-1)+1 \text{ if component } i \text{ is perceived faulty during execution } L \\ K \cdot \alpha_i(L-1) \text{ otherwise, with } K \in [0,1] \end{cases}$$

When $\alpha_i(L)$ becomes greater than or equal to a given threshold α_T, the damage to the state of component i is judged to be such that further diagnosis is necessary.

The appropriate filtering action can be provided by several different heuristics for the accumulation and decay processes (where $\alpha_i(L)$ takes slightly different expressions, see Chapter 4).

A distributed version of α-count is used in GUARDS to provide the error syndrome that is input to inter-channel fault diagnosis. Each channel i maintains C α-count variables: one representing its opinion of its own health (α_{ii}) and C-1 variables representing its opinions of the health of the other channels (α_{ij}, $j \neq i$). Since each channel may have a different perception of the errors created by other channels, the α-counts maintained by each channel must be viewed as single-source (private) values. They are thus consolidated through an interactive consistency protocol so that fault-free channels have a consistent view of the status of the instance. This then serves as the common input to a replicated fault diagnosis algorithm. The resulting diagnosis consists of a vector D whose elements D_i represent the diagnosed state of each channel (correct, or requiring passivation and isolation).

Once a channel has been diagnosed as requiring passivation, it is isolated (i.e., disconnected from the outside world) and reset (leading to re-initialisation of the operating system data structures). A thorough self-test is then carried out. If the test reveals a permanent fault, the channel is switched off and (possibly) undergoes repair. Whenever a channel passes the test (e.g., the fault was transient), or after having repaired a channel having suffered a permanent fault, it must be reintegrated to avoid unnecessary redundancy attrition.

The error filtering action of the α-count can be turned off by setting its threshold $\alpha_T = 1$. In this case, any transient fault leading to a self-detected error (or to errors perceived by a majority of channels) will cause that channel to go through the possibly lengthy self-test and re-integration procedure, irrespectively of the extent of the actual damage to the channel's state. A fault affecting another channel before re-integration of the former will induce a further decrease in the number of active channels. When transients are common, this policy can cause rapid switching to a safe state if the number of active channels becomes insufficient for error compensation to remain effective. The choice of whether filtering is used or, more generally, the value of the α-count threshold, thus leads to a classic trade-off between safety and reliability.

1.5.2 State Restoration

For a channel to be reintegrated, it must first resynchronise its clock, then its state, with the pool of active channels. Since not all state variables are necessarily consolidated through ICN exchanges, the state (or *channel context*) cannot be retrieved by simply observing the traffic on the ICN, but must be explicitly copied from the active channels. This is achieved by a system state restoration (SR) procedure applied to the channel context, i.e., the set of application state variables whose values are carried over successive iterations without consolidation.

A minimum level of service must be ensured, even during the SR procedure, so a limited number of vital application tasks must be allowed to continue execution on the active channels. We use an algorithm that we call *Running SR*. This is a multi-step algorithm where, at each step, only a fraction of the state is exchanged. Furthermore, vital application tasks may update state variables while SR progresses.

The basic behaviour of Running SR is the following. The channel context is arranged in a single (logical) memory block managed by a *context object*. When the state of the channel needs to be restored, it makes a *join* request to the pool of active channels. If the latter accept this request, the system enters an *SR mode*. The C-1 active channels enter a *put state* sub-mode while the joining channel enters a *get state* sub-mode.

To take advantage of the parallel links of the ICN, the whole block of memory storing the channel context is split into C-1 sub-blocks of similar size, each managed by one of the active channels. Each active channel i propagates to the joining channel any updates to state variables belonging to block i. A *Sweeper* task is executed to transfer the i-th block of the context. In the joining channel, transferred data are received and processed by a *Catcher* task. This task has most of the CPU time available since no application tasks are executed on that channel.

Switching from normal computation to the SR mode occurs at the beginning of an ICN frame, with a corresponding change in task scheduling, and SR completion always occurs at the end of a frame. After completion, signatures of the entire channel state are taken in each channel and exchanged through the interactive consistency protocol. State restoration is considered successful if all signatures match. Normal application scheduling is then re-activated on the next frame.

Since a deterministic, finite time is required to copy the memory block, and any updates to already copied state variables are immediately propagated, the whole (parallel) state restoration is performed in a deterministic, finite time. The state restoration tasks are assigned a priority and a deadline, and for schedulability analysis are treated the same as vital application tasks. Note that, during SR, the ICN has to support: a) the normal traffic generated by the vital (i.e., non-stoppable) applications, b) the extra traffic due to state variable updates, and c) the traffic generated by the Sweeper task. SR will therefore normally require a mode change to suspend non-vital application tasks so as to release processor time and ICN slots for SR execution and communication.

Diagnosis and state restoration are developed further in Chapter 4.

1.6 Output Data Consolidation

The purpose of the output data consolidation system (cf. Figure 1.1) is to map the replicated *logical outputs* of each channel onto the actual *physical outputs* to the controlled process, in such a way that the latter are either error-free or in a safe position. Such consolidation, placed at the physical interface with the controlled process, is the ultimate error confinement barrier, and is a complement to any software-implemented voting of the logical outputs.

A given instance of the architecture could have several different output consolidation mechanisms according to its various interfaces with the controlled process. Ideally, an output data consolidation mechanism should extend into the controlled process itself, to prevent the physical interface to the process from becoming a single point of failure. A typical example would be a control surface (e.g., in a fly-by-wire application) that can act as a physical voter by summing the forces produced by redundant actuators. Alternatively, a single channel can be designated to control a given actuator. Failures of that actuator can be detected at the application level by means of additional sensors allowing each channel to read back the controlled process variable and check it against the requested output. Recovery can then be achieved by switching to an alternative actuator. Other process-specific output data consolidation mechanisms used in the GUARDS end-user application domains include combinatorial logic implemented by relay or fluid valve networks, and the "arm-and-fire" technique commonly used to trigger space vehicle pyrotechnics (one channel sends an "arm" command, which is checked by the other channels, then all channels send matching "fire" commands; the pyrotechnics are triggered only if a majority of the latter concur with the former).

Output consolidation mechanisms such as these may be used for various end-user instances of the architecture. By definition, such process-specific techniques cannot be generic so no specific research has been carried out in this direction. However, the project has considered generic output consolidation mechanisms for networked and discrete outputs. A prototype consolidation mechanism has been implemented for discrete digital or analogue outputs that can be electrically isolated from each other and then connected through a wired-OR to the output devices. Consolidation is achieved by having each channel read back its own output and those of the other channel(s) so that a vote can be carried out in software. Each channel then sends selection signals to a hard-wired voter (one per channel) that enables or disconnects that channel's outputs. This approach relies on the assumption that the output devices can tolerate the short but inevitable output glitch caused by the read-back, vote and disconnection delay.

Chapter 5 gives further details regarding output consolidation.

1.7 Real-Time Scheduling

The architecture is capable of supporting a range of scheduling models (cf. Section 1.2.4) but we have chosen to focus on the standard pre-emptive priority-based scheme. Our timing analysis is based upon the Response-time Analysis [Leung & Whitehead 1982, Audsley *et al.* 1993]. We assume that any communication between applications is asynchronous through the shared memory. Furthermore, we assume the use of a non-blocking algorithm, such as that proposed in [Simpson 1990], to avoid the problems associated with remote blocking.

In this section, we consider the consequences of this choice of scheduling model on the coordination of replicated applications and on scheduling of the ICN network.

1.7.1 Inter-Channel Replication of Applications

For an application task to be replicated, it must behave deterministically and each replica task must process the same inputs in the same order. At any point where there is potential for replica divergence, the channels must perform an interactive consistency agreement. Unfortunately, the cost of executing interactive consistency agreement protocols can be significant. There is therefore a need to keep their use to a minimum.

With pre-emptive scheduling, there is no guarantee that replicas on different channels will read and write shared data items in the same order. To avoid having to resort to interactive consistency, we therefore force all replicated tasks to read the same internal data. This is achieved by means of a timestamp mechanism that ensures that all replicas read the same version of each data item. We can thus trade-off fewer agreement communications (and therefore greater efficiency) against early detection of errors. If we assume that each replica does not contain any inherently non-deterministic code, replica determinism and error masking (or detection) can be ensured by:

- Performing interactive consistency agreement or Byzantine agreement on single-sourced data.
- Ensuring that all replicas receive the same inputs when those inputs are obtained from other replica tasks (replicated inputs).
- Voting on any vital output.

1.7.2 Scheduling the ICN Network

Following the individual schedulability analysis of each channel, the following characteristics are known for each task participating in replicated transactions:

- Period
- Response-time
- Offset
- Deadline

The ICN tables can be built from this information in the same way as cyclic executive schedules can be constructed [Burns *et al.* 1995]. Since all communication through the channels' shared memory is asynchronous, the ICN manager can take the data any time after the producing task's deadline has expired.

Of course, there is a close relationship between the scheduling of the channels and the scheduling of the ICN network. If the off-line tool fails to find an ICN schedule, it is necessary to re-visit the design of the application.

Scheduling is detailed further in Chapter 3.

1.8 Architecture Development Environment

The generic architecture is supported by an Architecture Development Environment (ADE) [Paganone & Coppola 1997] consisting of a set of tools for designing instances of the architecture according to a coherent and rigorous design method. The tool-set allows collection of the performance attributes of the underlying execution environment and the analysis of the schedulability of hard real-time threads, not only within each processing element of the system, but also among them. This allows in particular a rigorous definition of critical communication and synchronisation among the redundant computers.

1.8.1 Design Method

The design and development of a GUARDS software application are based on a hard real-time (HRT) design method, which allows real-time requirements to be taken into account and verified during the design. The method also addresses the problem of designing replicated, fault-tolerant architectures, where a number of computing and communication boards interact for the consolidation of input values and output results.

The design of a GUARDS application is defined as a sequence of correlated activities, that may be re-iterated to produce a software design that complies with both the functional and non-functional requirements of the application. Three design activities are identified:

- *Functional Architecture design*, where the software application is defined through an appropriate design method.
- *Infrastructure Architecture design*, where the underlying computing environment of the application software is defined in terms of hardware boards and generic GUARDS software components.
- *Physical Architecture design*, where the functional architecture is mapped onto the infrastructure.

1.8.2 Supporting Tools

The Functional Architecture design should be supported by an appropriate method and tool. A survey and an analysis of design methods have shown that only HRT-HOOD [Burns & Wellings 1995b] addresses explicitly the design of *hard* real-time systems, providing means for the verification of their performance. Therefore, HRT-HOOD was selected as the baseline design method and HRT-HoodNICE adopted as supporting tool [Intecs-Sistemi 1996].

The Infrastructure Architecture design is supported by a specific tool-set that manages an archive of hardware and software components. Such components are described by their relations, compatibilities and performance attributes. The tool selects the needed components according to the characteristics of the required instance.

As part of the Physical Architecture design, the application tasks (i.e., HRT objects) identified in the functional architecture are mapped onto the Infrastructure Architecture. They are coupled with the real-time models of the selected components, in order to analyse and verify their schedulability properties. This is done by the Temporal Properties Analysis tool-set, which analyses the performance of the resulting distributed software system.

The Temporal Properties Analysis tool-set includes a Schedulability Analyser and a Scheduler Simulator, based on those available in HRT-HoodNICE. They have been enhanced to provide a more precise and realistic analysis (by taking into account the concept of thread offsets) and to cope with the specific needs of a redundant fault-tolerant architecture (by allowing the analysis of the interactions over the ICN).

A further result of the Physical Architecture design is that, on the basis of the real-time models produced by the verification tools, the critical interactions among software functions on different channels are scheduled in a deterministic way. The ICN transfer slots allocated to them and a set of pre-defined exchange tables are produced automatically.

As a final step of the design phase, the overall structure of the software application is extracted from the HRT-HOOD design and the related code is automatically generated. To this end, a set of mapping rules has been defined to translate the HRT-HOOD design in terms of threads implemented in a sequential programming language (which could be C or the sequential subset of Ada) and executed by a POSIX compliant microkernel.

A more detailed description of the Architecture Development Environment is given in Chapter 7.

1.9 Validation

The validation strategy implemented within GUARDS has two main objectives [Arlat 1997]:

- A short-term objective: the validation of the design principles of the generic architecture, including both real-time and dependability mechanisms.
- A long-term objective: the validation of the development of instances of the architecture implementing specific end-user requirements.

A large spectrum of methods, techniques and tools has been considered to address these validation objectives and to account for the validation requirements expressed by the recent IEC trans-application domain standard [IEC 61508].

Following the comprehensive development model described in [Laprie *et al.* 1995], the validation strategy is closely linked to the design solutions and the proposed generic architecture. The validation environment that supports the strategy includes components for verification and evaluation, using both analytical and experimental techniques. Figure 1.3 illustrates the relationship between the components of the validation environment, and their interactions with the architecture development environment.

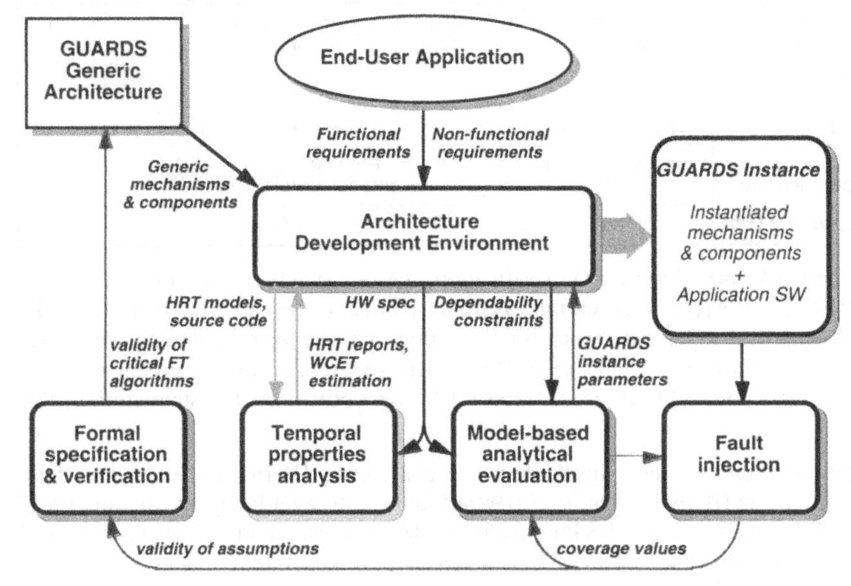

Figure 1.3 — Main Interactions between Architecture Development and Validation

Besides the three main validation components (namely, formal verification, model-based evaluation and fault injection), the figure explicitly identifies the role played by the schedulability analysis method and the supporting tool-set (cf. Section 1.8.2). The figure also depicts the complementarity and relationships among the three validation components. In particular, fault injection (carried out on prototypes) complements the other validation components by providing means for: a) assessing the validity of the necessary assumptions made during formal verification, and b) estimating the coverage parameters included in the analytical models for dependability evaluation. The following three subsections briefly describe the related validation activities.

1.9.1 Formal Verification

Formal approaches have been used for specifying and verifying critical dependability mechanisms. We concentrated our effort on four such mechanisms, which constitute the basic building blocks of the architecture: a) clock synchronisation, b) interactive consistency, c) fault diagnosis, and d) multi-level integrity.

The formal approaches that have been applied include both theorem-proving and model-checking. Table 1.1 summarises the main features of the verifications carried out for each of the target mechanisms.

Table 1.1 — Formal Verification Approaches

Target Mechanism	*Clock Synchronisation*	*Interactive Consistency*	*Fault Diagnosis*	*Multi-level Integrity*
Properties Verified	Agreement Accuracy	Agreement Validity	Correctness Completeness	Segregation Policy (Multi-level Objects)
Approach	Theorem-Proving	Model-Checking		
Description & Specification	Higher Order Logic	Process Algebra (CCS) and Temporal Logic (ACTL)		
Supporting Tool	PVS	JACK		

The work carried out on the verification of clock synchronisation relied heavily on PVS (Prototype Verification System) [Owre *et al.* 1996]. It led to the development of a general theory for averaging and non-averaging synchronisation algorithms [Schwier & von Henke 1997]. The verification of the synchronisation solution used in GUARDS (cf. Section 1.4.1) was derived as an instantiation of this general theory.

The verifications concerning interactive consistency [Bernardeschi *et al.* 1998a], fault diagnosis [Bernardeschi *et al.* 1998b] and multi-level integrity [Semini 1998] were all based on model-checking using the JACK (Just Another Concurrency Kit) tool-set [Bouali *et al.* 1994]. This integrated environment provides a set of verification tools that can be used separately or in combination. Due to the complexity of the required models, the tool-set was extended to include a symbolic model checker for ACTL [Fantechi *et al.* 1998].

Further details of formal verification of GUARDS mechanisms are given in Chapter 8.

1.9.2 Dependability Evaluation

Model-based dependability evaluation is widely recognised as a powerful means to make early and objective design decisions by assessing alternative architectural solutions. Nevertheless, fault-tolerant distributed systems (such as GUARDS instances) pose several practical modelling problems ("stiffness", combinatorial explosion, etc.). Moreover, due to the variety of the considered application domains, the dependability measures of interest encompass reliability, availability and safety.

To cope with these difficulties, we adopted a divide-and-conquer approach, where the modelling details and levels are tailored to fit the needs of the specific evaluation objectives. This was achieved by focusing first the modelling effort either on generic or specific architectural features, or on selected dependability mechanisms. Then, an abstract modelling viewpoint was devised. This aims to provide a global framework for configuring instances to meet specific application dependability requirements. Finally, elaborating on previous related work (e.g., [Kanoun *et al.* 1999]), we investigated a detailed modelling viewpoint that supports incremental and hierarchical evaluation.

Table 1.2 identifies the various dependability evaluation activities carried out according to these three modelling viewpoints.

Table 1.2 — Dependability Evaluation Viewpoints and Studies

Modelling Level	*Focused*	*Abstract*	*Detailed*
Targeted Mechanisms, Strategies, Instances	- α-count mechanism - Phased missions - Intra-channel error detection mechanism	- Railway prototype Instance - Nuclear submarine prototype Instance - Space prototype instance	Overall design and interactions (Nuclear submarine prototype instance)
Formalism	Stochastic activity networks and generalised stochastic Petri nets	Generalised stochastic Petri nets	Stochastic Petri nets
Supporting Tools	UltraSAN, SURF-2	SURF-2	MOCA-PN
Resolution Methods	Analytical and Monte-Carlo simulation	Analytical and method of stages	Monte-Carlo simulation

The focused models address several issues concerning the analysis of generic mechanisms (e.g., α-count [Bondavalli *et al.* 1997a]) and of specific features for selected instances (phased missions, for the space prototype instance [Bondavalli *et al.* 1997b], intra-channel error detection for the railway prototype instance).

The second viewpoint aims to establish a baseline set of models of some domain-specific instances of the architecture [Powell et al. 1998b]. A general notation is introduced that allows for a consistent interpretation of the model parameters (layers, correlated faults, etc.) for each prototype instance. This work provides the foundation of a generic modelling approach to guide the choice of a particular instantiation of the architecture, according to the dependability requirements of the end-user application. A large number of parameters (proportion of transient vs. permanent faults, correlated faults in the hardware and software layers, coverage factors, error processing rates, etc.) have been included in the models, allowing intensive sensitivity analyses to be carried out.

Detailed models are needed to allow for a more comprehensive analysis of the behaviour of the instances (dependencies, error propagation, etc.). Specific work has addressed hierarchical modelling with the aim of mastering the complexity of such detailed models [Jenn 1998a]. This work is directed mainly at: a) enforcing the thoroughness of the analysis, b) helping the analyst (i.e., a design engineer who is not necessarily a modelling expert).

Although they were supported by different tools, namely UltraSAN [Sanders *et al.* 1995], MOCA-PN [Dutuit *et al.* 1997] and SURF-2 [Béounes *et al.* 1993], the modelling efforts all rely on the stochastic Petri net formalism. This facilitates re-use of the models and modelling methodology according to the various viewpoints.

Further details concerning the dependability evaluation studies carried out in GUARDS are given in Chapter 9.

1.9.3 Fault Injection

The main objectives of fault injection within the overall validation strategy are twofold: a) to complement the formal verification of GUARDS mechanisms (i.e., removal of residual deficiencies in the mechanisms), and b) to support the development of GUARDS instances by assessing their overall behaviour in the presence of faults, in particular by estimating coverage and latency figures for the built-in error detection mechanisms [Arlat *et al.* 1990].

Indeed, as an experimental approach, fault injection provides a pragmatic means to complement formal verification by overcoming some of the behavioural and structural abstractions made, especially regarding the failure mode assumptions. Fault injection is carried out on complete prototypes so the mechanisms are tested globally when they have been integrated into an instance. In particular, the interactions between the hardware and software features are taken into account.

Although available tools could have been used (albeit with some extensions), a specific fault injection tool-set (FITS) has been developed to support the end-users in the development of specific instances of the generic architecture.

Both for cost-effectiveness and flexibility, the fault injection environment is based on the software-implemented fault injection (SWIFI) technique [Hsueh *et al.* 1997]. This also allows tests to be conducted more efficiently, since: a) a limited number of errors can simulate the consequences of a large number of faults, b) it is less likely that the injected error fails to exercise the dependability mechanisms.

Two main levels of injection are considered, according to whether the targeted mechanisms are implemented by the ICN-manager board or by the intra-channel processors. In practice, the implementations differ significantly: whereas fault injection on the intra-channel processors can be assisted by the resident COTS operating systems and debug facilities [Carreira *et al.* 1998, Krishnamurthy *et al.* 1998], the ICN-manager has only a very simple cyclic executive, so specific communication and observation facilities had to be implemented.

Besides injecting specific fault/error types, FITS allows injection to be synchronised with the target system by monitoring trigger events. Of course, the observations depend on the targeted mechanisms. While it is primarily intended to inject on a single channel, observations are carried out on all channels.

Further details on FITS may be found in [Oswald & Attermeyer 1999, Oswald *et al.* 1999].

1.10 Domain-Specific Instances of the Architecture

Several practical instances of the generic architecture have been studied. The configurations of the instances are quite different, as are their fault-tolerance strategies. Moreover, although the considered operating systems are POSIX-compliant, they are not identical, neither are the end-users' preferred system development environments. Consequently, although there is a single specification of the generic software components of the fault-tolerant and integrity management layer, they have different practical instantiations in each instance.

1.10.1 Railway Instances

One instance studied for the railway domain is a fairly classic triple modular redundant (TMR) architecture with one processor per channel (Figure 1.4). It features Motorola 68040 or 68360 processors, each running a POSIX-compliant VxWorks operating system.

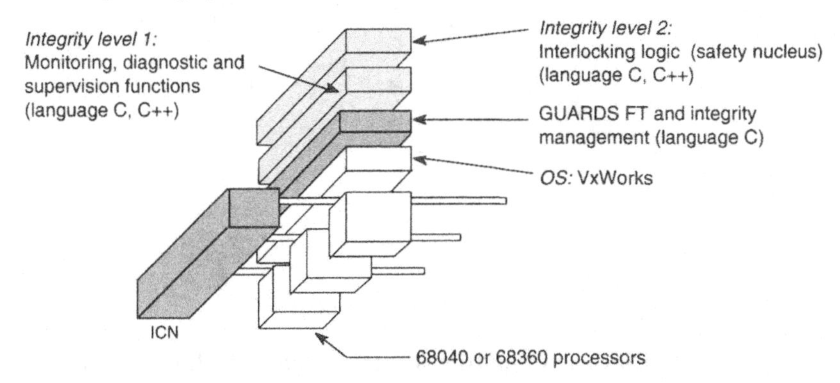

Figure 1.4 — Railway Triplex Instance (*C*=3, *L*=1, *I*=2)

Compared to currently-deployed systems, the innovative aspect of this architecture is the co-existence of two levels of application software of very different degrees of criticality:

- Highly-critical interlocking logic or safety nucleus, which is at the highest integrity level.
- Monitoring, diagnostic and supervision functions, which are of the lowest criticality.

This is a significant departure from current practice in railway applications, where these two levels of integrity would normally be implemented on separate instances. However, there is an appreciable economic advantage to be gained when it is possible to share the same hardware between both levels (e.g., for small railway stations).

As long as there are three operational channels, any errors due to a single faulty channel are assumed to be masked (and detected) by majority voting. While there are only two operational channels, the instance operates in a two-out-of-two mode.

Should a fault occur while in this mode, the instance is switched to a safe state if the errors caused by the fault are detected (either locally within a channel or by two-out-of-two comparison).

A second railway instance was studied and actually implemented as a demonstrator. This is a fairly straightforward duplex fail-safe configuration, which is described in Section 10.3 (Chapter 10).

1.10.2 Nuclear Propulsion Instance

The targeted nuclear propulsion application is a secondary protection system. The considered instance a dual-channel architecture with two Pentium processors in each channel (Figure 1.5). To prevent common-mode failures of both channels due to physical damage, the channels are geographically separated one from the other by a distance of 10 to 20 meters. Like the railway application, this instance hosts two levels of integrity.

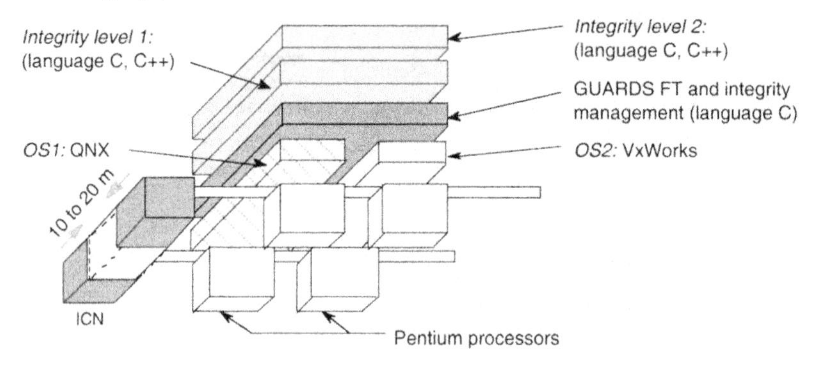

Figure 1.5 — Nuclear Propulsion Target Instance (*C*=2, *L*=2, *I*=2)

An innovative aspect of this prototype is the use of two processors (or *lanes*) in each channel, with two different POSIX-compliant operating systems: QNX and VxWorks. Apart from the operating systems, both lanes run identical software to implement duplication-and-comparison within each channel. The aim is to be able to detect errors due to design faults in these COTS operating systems.

All application software components are executed on both processors in both channels. Within a channel, the copies of an application component on lanes 1 and 2 are provided with the same inputs in the same order. In the absence of faults, both copies should provide identical results. These are compared on a bit-to-bit basis. All application components within a channel are thus configured as self-checking pairs to provide detection of errors due to faults activated independently in each lane.

In particular, the covered fault classes include physical faults (of the processors) and design faults of the processors and their operating systems. Note that, in this instance, an assumption of independent activations for design faults of the operating systems can be based on the fact that their designs are diversified. For design faults of the processors, an assumption of independent activation can be based on the decorrelation of their utilisation conditions (due to loose coupling and diversification of operating systems).

As long as both channels are operational, they operate in a two-out-of-two mode. Results of computations that are declared as error-free by the intra-channel mechanisms are compared and, in case of disagreement, the instance is put into a safe state. However, if errors are detected locally, by intra-channel mechanisms, the channel declares itself to be faulty and the instance switches to single channel operation. Note that this strategy is different to that of the two-channel configurations of the railway instances (duplex instance, or triplex instance degraded to duplex); those instances switch to a safe state whether the error is detected locally or by comparison.

The implementation of a demonstrator of this instance is described in Section 10.4 (Chapter 10).

1.10.3 Space Instance

The instance considered for space applications is a full four-channel instance capable of tolerating arbitrary faults at the inter-channel level (Figure 1.6). Degradation to three-, two- and one-channel operation is possible. In common with the railway and nuclear applications, this instance also features two levels of integrity.

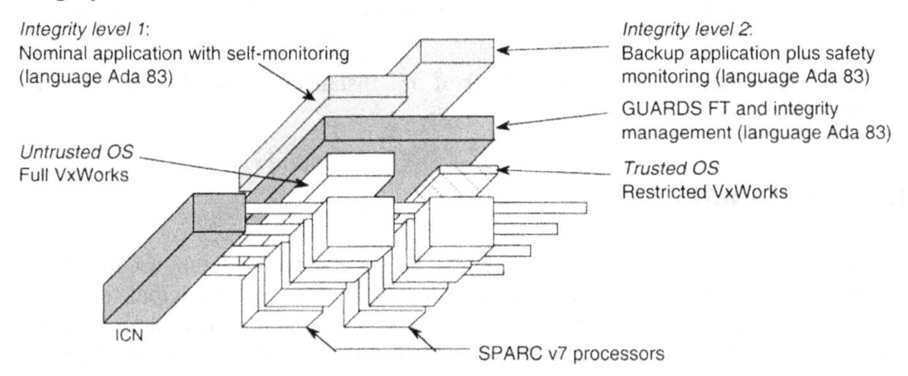

Figure 1.6 — Space Target Instance (C=4, L=2, I=2)

Like the prototype for the nuclear propulsion application, this instance also possesses two lanes, but for a different reason. For the nuclear application described in the last section, the aim was to allow diversified but equivalent operating systems to be used so that errors due to design faults could be detected. Here, the objective is to have one of the lanes act as a back-up for the other lane.

We refer to the two lanes as the *primary* and *secondary* lanes. Each lane supports a different operating system and different application software:

- The primary lane runs a full-functionality version of VxWorks and a *nominal* application that provides full control of the spacecraft and its payload. The application includes built-in self-monitoring based on executable assertions and timing checks.
- The secondary lane runs a much simpler, restricted version of VxWorks and either a safety-monitoring application or a simple back-up application. The

purpose of the latter is to provide control of the spacecraft in a very limited "survival" mode (e.g., sun-pointing and telemetry/telecommand functions).

The idea is that neither the full VxWorks nor the nominal application supported by the primary lane can be trusted to be free of design faults. However, the restricted version of VxWorks and the application software supported by the back-up lane are assumed to be free of design faults and thus trustable. The aim is to allow continued (but severely-degraded) operation in the face of a correlated fault across all processors of the primary lane. Errors due to such a correlated fault can be detected in two ways:

- Self-monitoring functions included within the nominal application.
- A safety-monitoring application executed by the secondary lane while the primary lane is operational.

In view of the differing levels of trust of the applications supported by the primary and secondary lanes, they are placed at different levels of integrity. The nominal application (on the primary lane) is not trusted, so it is assigned to integrity level one. The back-up application is assumed to be free of design faults and is placed at integrity level two. This separation of the integrity levels on different lanes provides improved segregation ("fire-walling") between the two levels of integrity.

The implementation of a demonstrator of this instance is described in Section 10.2 (Chapter 10).

1.11 Summary

This chapter has outlined the motivations for the generic fault-tolerant architecture and its associated development and validation environment. The principal features of the architecture have been briefly described and the main validation activities delineated. We have also described the domain-specific instances of the architecture that have guided the thought process of the project team. Each of these aspects is described in more detail in the remainder of the book.

Chapter 2
Inter-Channel Communication Network

Central to the architecture is an inter-channel communication network (ICN), which consists of an ICN-manager for each channel and unidirectional serial links to interconnect the ICN-managers. In the current implementation, which is detailed in Section 10.1 (Chapter 10), the ICN-manager is a Motorola 68040-based board with a dual-port shared memory for asynchronous communication with the intra-channel VME back-plane bus. Serial links are provided by Motorola 68360-based piggy-back boards. Each such board provides two Ethernet links. One link is configured as transmit only, the other links are configured as receive only. An ICN-manager can thus simultaneously transmit data to the remote ICN-managers over its outgoing serial link and receive data from the remote ICN-managers over the other links.

This inter-channel dimension is the primary fault containment region of the architecture. To allow GUARDS-based architectures to reach very high levels of dependability, all inter-channel exchanges must take into account arbitrary faults. In this context, the two essential functions that the ICN must provide in a fault-tolerant way are:

- Synchronisation of the different channels, through the construction of a global clock.
- Exchange of data between channels, with interactive consistency (consensus) on non-replicated data.

Sections 2.1 and 2.2 of this chapter describe, for each of these two functions, the rationale behind the choice of a particular solution and detail the one that has been adopted. Section 2.3 then focuses on the way inter-channel messages are authenticated, taking into account the tight performance constraints of a real-time system.

2.1 Clock Synchronisation

This section first presents some commonly-used definitions and notations for clock synchronisation, and then gives an overview of the known solutions that meet the

D. Powell (ed.), A Generic Fault-Tolerant Architecture for Real-Time Dependable Systems, 27–50.

GUARDS constraints. The adopted synchronisation algorithm is then detailed, including the way by which initial synchronisation is achieved.

2.1.1 Definitions and Notations

We consider a set of fully-connected channels or *nodes*. Each node has a physical clock and computes a logical clock time.

Definition 1: Clock Time

Each node maintains a logical clock time $T = C(t)$, meaning that at real-time t the clock time of the local node is T.

By convention, variables associated with clock time (local to a given node) are in uppercase, whereas variables associated with real time (measured in an assumed global Newtonian frame) are in lowercase.

Definition 2: Drift Rate

A non-faulty physical clock $C_i(t)$ has a maximum drift rate ρ if and only if for any $t_2 > t_1$:

$$1-\rho < \frac{C_i(t_2) - C_i(t_1)}{t_2 - t_1} < 1+\rho$$

Definition 3: Agreement

The agreement condition is satisfied if and only if the skew δ between any non-faulty logical clocks $C_i(t)$ and $C_j(t)$ is bounded:

$$\left| C_i(t) - C_j(t) \right| \leq \delta$$

Definition 4: Accuracy

The accuracy condition is satisfied if and only if any non-faulty logical clock $C_i(t)$ stays within a linear envelope of real time, i.e., there exists a constant $\gamma > 0$ and two constants a and b (that depend on initial conditions) such that:

$$(1-\gamma)t + a \leq C_i(t) \leq (1+\gamma)t + b$$

Definition 5: Synchronisation Algorithm

A synchronisation algorithm is an algorithm that satisfies the agreement and accuracy conditions. Clock synchronisation is generally achieved through repeated resynchronisation phases. Let R be the interval between two such resynchronisations.

Some typical values of the parameters introduced so far or used in the remainder of the section are given in Table 2.1.

Note that the maximum skew between any two logical clocks is necessarily greater than the skew between their physical clocks, and that this latter skew can easily be obtained through the simple formula $\delta_{physical} = 2\rho R$. With the values given in Table 2.1, this gives $\delta_{physical} = 2\rho R = 2 \cdot 10^{-4} \cdot 500\text{ms} = 100\mu\text{s}$, which in fact leads to a maximum skew, δ, slightly greater than 100μs (see Section 2.1.2.3 for a more precise estimation of δ).

Table 2.1 — Typical Values of Clock Synchronisation Parameters

Variable	Content	Typical value
ρ	Maximum drift rate of all non-faulty physical clocks	10-4 sec/sec
f	Quartz frequency of physical clocks	10 MHz
δ	Maximum skew between any two non-faulty logical clocks	100 μs
d	Upper bound on the transmission delay	50 μs
ε	Upper bound on the read error[1]	1 μs
n	Number of clocks	≤4
m	Maximum number of (arbitrarily) faulty clocks	1
R	Resynchronisation interval	500 ms

2.1.2 Existing Algorithms

Many clock synchronisation algorithms are described in the literature. [Ramanathan *et al.* 1990] provides a good survey of these algorithms. In line with our requirement to minimise the amount of specific hardware, we restrict ourselves to software-implemented algorithms. This yields three main kinds of algorithms (Figure 2.1):

- *Convergence algorithms*, where nodes periodically resynchronise by exchanging synchronisation messages.
- *Consistency algorithms*, where nodes use an interactive consistency algorithm to achieve agreement on clock values. We chose not to consider these since there exists only a limited amount of literature on this kind of algorithm [Lamport & Melliar-Smith 1985].
- *Probabilistic algorithms* [Cristian 1989], where nodes can make the worst-case skew as small as desired, but with an increasing probability of loss of synchronisation. This is clearly unacceptable in the field of safety-critical applications.

Focusing then on convergence algorithms, we observe that there are two main kinds:

- *Convergence-averaging algorithms*, where each node resynchronises according to clock values obtained through periodic one-round clock exchanges. On each node, the other clocks can be taken into account through

[1] The read error is due to the transmission delay uncertainty, i.e., $d - 2\varepsilon$ is the minimum message transit delay between any two nodes in the system.

a mean-like function [Lamport & Melliar-Smith 1985], or a median-like function [Lundelius-Welch & Lynch 1988].

- *Convergence-nonaveraging algorithms*, where each node periodically seeks to be the system synchroniser. To deal with possible Byzantine behaviour, the exchanged messages can be authenticated or broadcast [Srikanth & Toueg 1987, Dolev *et al.* 1995].

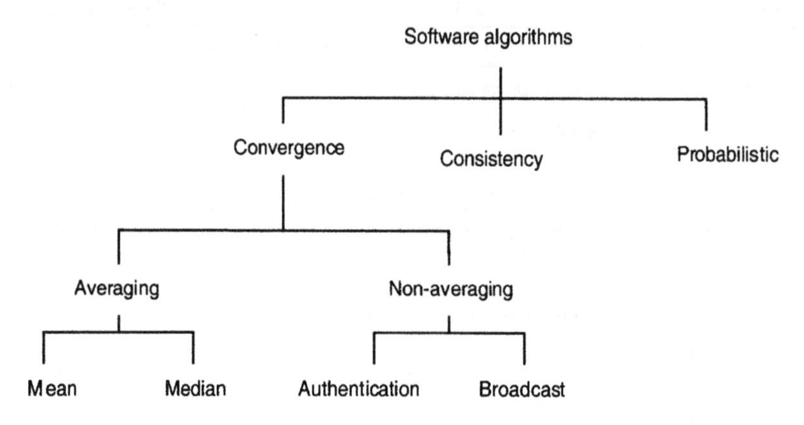

Figure 2.1 — Synchronisation Algorithms

2.1.2.1 Convergence-averaging Algorithms

These synchronisation algorithms can tolerate m arbitrary faulty nodes in a (fully connected) network of n nodes, under the necessary condition, even with authentication, that $n \geq 3m+1$. This implies in particular that they *cannot* tolerate a Byzantine faulty clock in a three-node system.

The basic idea of these algorithms is as follows. The resynchronisation is performed periodically: each node broadcasts a resynchronisation message when its local clock time has counted R seconds since the last resynchronisation period. At the same time, the node collects during a given waiting period (which is far shorter than R) the resynchronisation messages broadcast by other nodes: it records the arrival time (according to its local clock time) of each received resynchronisation message. After the waiting period, it uses a fault-tolerant averaging function F on the arrival times to compute an averaged clock time:

- In [Lamport & Melliar-Smith 1985], the fault-tolerant averaging function F is the mean of all values (with values differing from the local value by more than δ replaced by the local value).

- In [Lundelius-Welch & Lynch 1988], the function F is a median-like function. It consists in rejecting the m highest and m lowest values, and then performing the average of all remaining values.

Each node then corrects its own clock time by setting it to the averaged clock time before the next resynchronisation period.

Due to its simpler fault-tolerant averaging function, we choose the algorithm of [Lundelius-Welch & Lynch 1988] to represent the class of convergence-averaging algorithms in the remainder of this chapter.

With three clocks, the fault-tolerant averaging function of [Lundelius-Welch & Lynch 1988] reduces to a true median. If a Byzantine faulty clock sends synchronisation messages to two other (non-faulty) clocks such that the earliest (respectively latest) non-faulty clock receives the Byzantine synchronisation message before (respectively after) its own synchronisation emission (see Figure 2.2), then the non-faulty clocks both find themselves as being equal to the calculated median. They thus retain their own values and drift progressively apart.

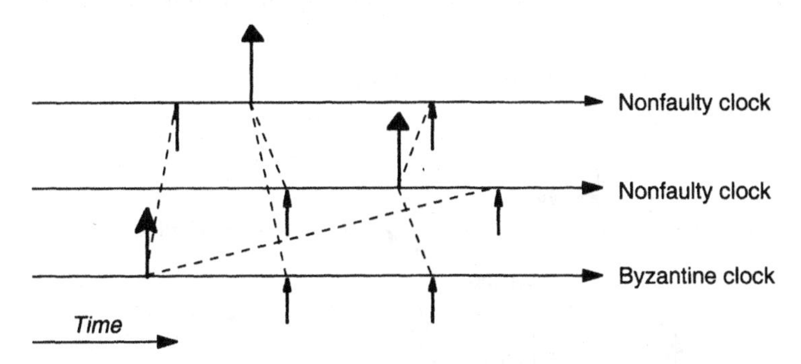

Figure 2.2 — Desynchronisation of Two Non-faulty Clocks in Presence of a Byzantine Clock

2.1.2.2 Convergence-nonaveraging Algorithms

When authentication is used for inter-node message exchanges, these synchronisation algorithms can tolerate m arbitrary faulty node with only $n \geq 2m+1$ nodes. This implies in particular that they *can* tolerate a Byzantine faulty clock in a three-node system.

In these algorithms [Srikanth & Toueg 1987, Dolev *et al.* 1995], each node periodically seeks to be the system synchroniser, and succeeds if it is the fastest non-faulty node. More precisely, each node broadcasts its resynchronisation message and collects during a given waiting period the resynchronisation messages broadcast by other nodes (like in convergence-averaging algorithms). As soon as it has received $m+1$ resynchronisation messages (including possibly its own), it restarts its local clock for the next period and relays the $m+1$ resynchronisation messages to all other nodes. It should be noted that the worst case skew of all such algorithms proposed so far is greater than the transmission delay: $\delta > d$.

To better understand the mechanisms involved here, we will detail the algorithm proposed in [Srikanth & Toueg 1987]. In this algorithm, each node starts a new local clock $C^k(t)$ at each successive resynchronisation round. The core of the algorithm can be described by the following pseudo-code (the constant a being chosen so that clock-time does not go backward when resynchronisation occurs, i.e., $C^k(t) > C^{k-1}(t)$:

```
if C^{k-1}(t) = kR
        then    sign and broadcast (round k)

//

if (m+1) signed (round k) messages received
        then    C^k(t) = kR + a
                Relay all (m+1) messages to all
```

When a node broadcasts its synchronisation message, it is assumed that it receives it immediately (because for the local node, there is no transmission delay for the reception of its own synchronisation message). So, the test of reception of $m+1$ signed messages includes the possible locally emitted synchronisation message.

Figure 2.3 gives an example of tolerance of a Byzantine faulty clock in a three node system (m is equal to 1).

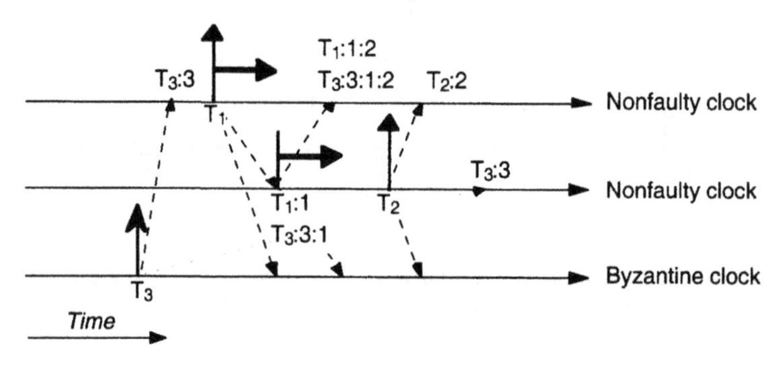

(signature of a message *m* by a node *i* is denoted m:i)

Figure 2.3 — Tolerance of a Byzantine Clock in a 3-Node System

The faulty clock (clock number 3 on the figure) broadcasts its (signed) resynchronisation message so that the fastest non-faulty clock (clock number 1 on the figure) receives this message before the end of its own resynchronisation interval and the slowest non-faulty clock (clock number 2 on the figure) receives it after the end of its own resynchronisation interval. We have seen that such an inconsistent broadcast can desynchronise the two non-faulty clocks under a convergence-averaging synchronisation algorithm. Let us analyse what happens here:

- The fastest non-faulty clock receives the resynchronisation message of clock 3. A short time after, this non-faulty clock reaches the end of its own resynchronisation interval. According to the algorithm, it has received at this time $m+1=2$ resynchronisation messages (the one coming from clock 3 and its own). So it starts its new local clock (horizontal bold arrow on the figure) and relays to the two other nodes two (signed) resynchronisation messages.

- The slowest non-faulty clock receives the two signed resynchronisation messages (just emitted by the clock 1), and thus according to the algorithm restarts its new local clock (horizontal bold arrow on the figure) and relays to the two other nodes these two (signed) resynchronisation messages. A short time after, this clock 2 reaches the end of its own resynchronisation interval and then broadcast its own resynchronisation message. A short time after, this clock 2 receives the resynchronisation message from clock 3, which it may identify as being faulty.

We can see that the clock 2 starts its new local clock at most d seconds (maximum transmission time) after clock 1 has started its own.

2.1.2.3 Maximum Skew Estimation

One of the most important characteristics of a synchronisation algorithm is its ability to achieve a small maximum skew between non-faulty clocks. Therefore, it is important that we estimate and compare the maximum skew of [Lundelius-Welch & Lynch 1988] (hereafter denoted by LL) and [Srikanth & Toueg 1987] (hereafter denoted by ST).

Theoretical results provide us with the two following formulas:

- For the LL convergence-averaging algorithm [Lundelius-Welch & Lynch 1988, theorem 16, p.21][2]:

$$\delta = \beta + \varepsilon + \rho(7\beta + 3d + 4\varepsilon) + 4\rho^2(2 + \rho)(\beta + d)$$

  with $\beta = 4(\varepsilon + \rho R)$

- For the ST convergence-nonaveraging algorithm [Srikanth & Toueg 1987, lemmas 6 and 8, pp.630-1]:

$$\delta = \left(R(1+\rho) + d\right)\left(\frac{\rho(2+\rho)}{1+\rho}\right) + d(1+\rho)$$

Using the typical values given in Table 2.1, we have $\varepsilon = 10^{-6}\mathrm{s}$, $\rho R = 5 \cdot 10^{-5}\mathrm{s}$, $d = 5 \cdot 10^{-5}\mathrm{s}$, and we can see that terms including factors like ρ^2, $\rho\varepsilon$ and ρd are negligible. By dropping these higher order terms, we obtain:

- For the LL algorithm: $\delta \approx 5\varepsilon + 4\rho R$
- For the ST algorithm: $\delta \approx d + 2\rho R$

We can see that the maximum skew δ is of the same order of magnitude in both algorithms (a few hundreds of microseconds). However, several remarks must be made:

- We have taken a rather pessimistic maximum drift rate of 10^{-4} for physical clocks. For example, ρ is estimated at 10^{-6} by [Ramanathan *et al.* 1990], leading to $\rho R = 5 \cdot 10^{-7}\mathrm{s}$. If we take such a small value for ρ, then LL becomes better than ST.

---

[2] The factorisation of this fomula follows that given in [Ramanathan *et al.* 1990]. But note that the exact reformulation of [Ramanathan *et al.* 1990] is incorrect (omission of the scalar 3 in front of *d* in the third term).

- We have taken a rather optimistic maximum transmission delay *d*, taking into account that GUARDS channels are expected to be near each other (typically a few decimeters). If channels are distributed over longer distances, then this could lead to an increase of *d*, thus making again LL better than ST.

- From a practical viewpoint, it is not clear whether the transmission delay *d* is the same for LL and for ST. Indeed, LL is a single-round algorithm where synchronisation messages are deterministically and synchronously broadcast. Thus *d* corresponds to the maximum time required for a message to be prepared by a node, broadcast to all other nodes and processed by these other nodes. For ST the issue is that a node may have to broadcast its synchronisation message (first round), and then to relay immediately the synchronisation messages it received (second round), precisely because it received the $(m+1)^{th}$ signed synchronisation message just after having sent its own synchronisation message. The maximum transmission delay has to take into account this unfavourable case (which is likely to happen precisely when clocks are well synchronised), thus leading to a larger value.

2.1.3 GUARDS Synchronisation Algorithm

At this stage, we are in the position to choose between the two main candidate algorithms: LL or ST. The actual choice was made after a carefully-conducted technical trade-off.

2.1.3.1 Technical Trade-off

A given GUARDS instance is composed of at most four channels, fully connected together through the Inter-Channel Network (ICN). For the sake of simplicity, we assume that each channel contains only one node (the problem of synchronising all the nodes inside a given channel is another issue, which we do not analyse here). We thus have $n = C \leq 4$.

Building on Section 2.1.2 we consider only convergence algorithms. These algorithms are based on a periodic resynchronisation interval.

We define an active node as a node that does not consider itself to be faulty (A node which considers itself as faulty does not execute the synchronisation algorithm, but reconfigures itself and then tries to reintegrate the pool.). Let *n* be the number of active nodes. If we make no assumption about the ICN, then we have to deal with possible Byzantine faulty clocks:

- When $n = 4$, then both convergence-averaging or convergence-nonaveraging algorithms can be used.

- When $n = 3$, only a convergence-nonaveraging algorithm can be used. But this solution may lead to a maximum skew (greater than the transmission delay) that is too large (see Section 2.1.2.3).

The above considerations are for the general case, where no assumptions are made on the way the nodes are interconnected together, apart from the fact that they are indeed fully interconnected. However, in GUARDS, we deliberately chose to implement the ICN using a broadcast topology (see Figure 2.4). We can therefore practically exclude the case of a Byzantine clock, and thus consider only convergence-averaging algorithms. Indeed, when a given node sends a message, the message is physically broadcast to all other nodes by hardware. Thus, for each emission, Byzantine faults can occur only during transmission or reception of messages. Moreover, such faults can only be value faults (corrupted messages), and not timing faults: if a message is emitted by a node, then on each remote node it is either received within the upper bound on transmission delay (whatever the value is) or not received at all.

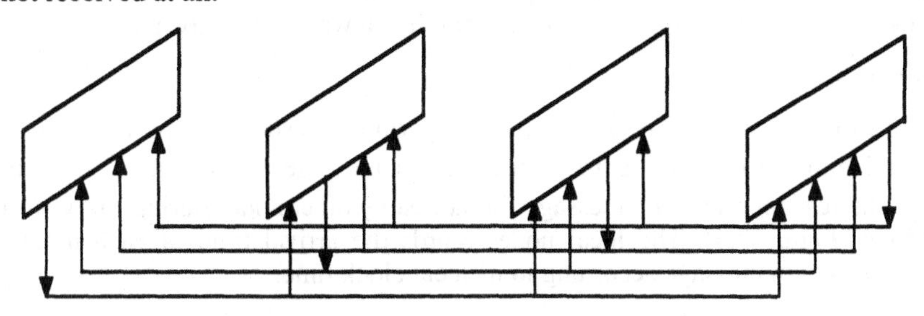

Figure 2.4 — ICN Broadcast Topology

However, it is still possible to imagine particular cases (very unlikely, but not theoretically impossible), where the existence of a broadcast network does not totally preclude the possibility of a Byzantine clock:

- One case where a Byzantine faulty clock could still be observed would be a faulty node which periodically sends two successive synchronisation messages slightly degraded in such a way that these two messages are received only by one remote node each (as in Figure 2.2). For example, the two remote nodes have different physical receivers, so some bits of the first emitted synchronisation message could be degraded and received only by one remote node, and some other bits of the second emitted synchronisation message could be degraded and received only by the other remote node.
- Another case, perhaps less intricate, would be to consider a decrease in the transmission speed of one link between its broadcast point and one of the receivers. In such a case, the concerned receiver may receive the emitted message significantly later than the other receivers, thus allowing a case like in Figure 2.2 to occur.

Briefly, the technical trade-off is the following:

- Assume the possible presence of Byzantine clocks. This leads to synchronisation algorithms involving several rounds of message exchange (dynamically managed at each cycle) and having a large skew between clocks (which could impede the genericity of GUARDS for some demanding applications).
- Assume that there are no Byzantine clocks when there are only three channels, thanks to the broadcast nature of the ICN network. This allows the choice of a simple (static one-round message exchanges scheme) and efficient (small skew) algorithm.

We have thus chosen the second alternative. This implies that, in the case $n = 3$, the probability of occurrence of a Byzantine clock (expected to be extremely small) should be considered for very critical applications.

2.1.3.2 The GUARDS Algorithm

The algorithm actually implemented in GUARDS is a convergence-averaging algorithm, with a fault-tolerant averaging function F that depends on the number of active nodes (it is actually the LL algorithm applied to the four-node case). It is composed of successive steps (described from the viewpoint of a given node).

Reception Step

Let T_k be the end of the kth resynchronisation period (measured on the local clock). The local node broadcasts its resynchronisation message at clock time T_k, and collects the resynchronisation messages broadcast by other nodes during the waiting period (see Figure 2.5). The local node records the arrival time of each received resynchronisation message according to its local clock time.

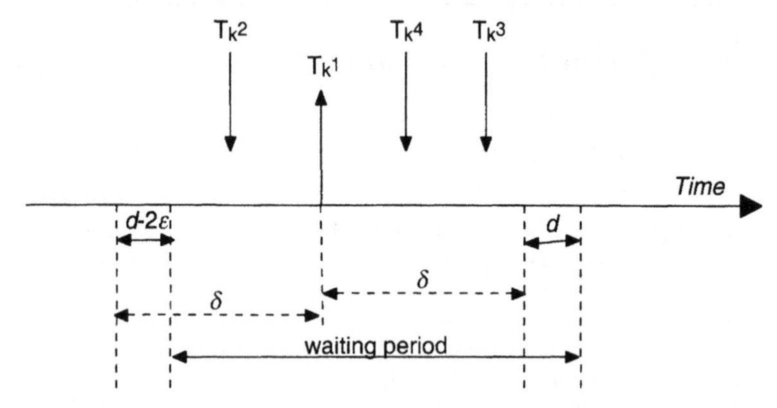

Figure 2.5 — Waiting Period

The waiting period should be of length 2δ, and would be centred on T_k if transmissions were instantaneous. To take into account the minimum (resp. maximum) transmission delay, the beginning (resp. end) of the waiting period is in fact shifted by $d-2\varepsilon$ (resp. d) into the future. Thus the exact waiting period is $\left[T_k - \delta + d - 2\varepsilon ; T_k + \delta + d\right]$.

Computing Step

Let j be the number of received values (including the local value) and n be the number of currently active nodes. Since we aim to tolerate only non-simultaneous faults, we have either $j = n+1$ (one node is joining), $j = n$ or $j = n-1$ (one node is missing)[3].

---

[3] Note that when $j = n$, it may be the (improbable) case that one node is joining and simultaneously one node is missing.

Then, the averaged clock time T is computed as follows:

- For $j = 4$, T is the mean of the two middle values.
  This corresponds exactly to the fault-tolerating function used in [Lundelius-Welch & Lynch 1988]. Moreover, n is necessarily equal to 4, unless one node is joining (in which case $n = 3$ and will be set to 4 at the next cycle). If one value is at a distance strictly greater than δ from any other value (there is at most one such value, since we preclude the cases of simultaneous faults), then the corresponding node is suspected to be faulty[4]. If it is a joining node that is suspected to be faulty, then it is not integrated in the pool and n remains equal to 3 (a similar remark applies in the following cases).
- For $j = 3$, T is the middle value.
  If $n = 4$, the node corresponding to the missing value is suspected to be faulty. If $n = 3$ or $n = 2$ (one node is joining) and one value is at a distance strictly greater than δ from any other value, then the corresponding node is suspected to be faulty.
- For $j = 2$, T is the mean of the two values.
  If $n = 3$, the node corresponding to the missing value is suspected to be faulty. If $n = 2$ or $n = 1$ (one node is joining) and if the two values are at a distance strictly greater than δ, then both nodes are suspected to be faulty.
- For $j = 1$, T is the only received value (it is necessarily the local value).
  If $n = 2$, then both nodes are suspected to be faulty[5]. If $n = 1$, then just proceed. The case $n = 0$ is vacuous (one node never joins itself).

Correction Step

The correction to apply to the length of the next resynchronisation period is $\Delta = T - T_k$:

- If Δ is positive, then the local node is early.
- If Δ is negative, then the local node is late.

In both cases, the duration of the next resynchronisation period of the local node is set to $R + \Delta$ seconds, i.e., $T_{k+1} = T_k + R + \Delta = T + R$.

2.1.4 Initial Synchronisation

Two solutions can be distinguished to perform initial synchronisation of a set of nodes:

- Start all nodes independently. Initial synchronisation algorithms exist that can synchronise such a set of non-synchronised active nodes (even in the

---

[4] This status will be subsequently used to effectively identify the faulty node. We focus here just on the synchronisation algorithm, without taking into account the pool reconfiguration mechanism.

[5] At this stage it is impossible to decide which of the two nodes is actually faulty. If reconfiguration from two channels to one channel is required, then further diagnosis is necessary, e.g., through autotest.

presence of arbitrary failures) [Srikanth & Toueg 1987, Lundelius-Welch & Lynch 1988].

- Ensure that the nodes are started one after another. In this case, the problem of initial synchronisation is reduced to the problem of joining a set of already synchronised nodes (assuming that no fault occurs during the execution of the joining algorithm).

The second solution is far simpler than the first, and is needed anyway (to allow a failed node to be recovered and re-integrated into the set of active nodes). Thus, it is this solution that has been implemented.

When a node is started, it considers that it is started at the beginning of its local resynchronisation period. It sets the duration of its first resynchronisation period, R_1, equal to $3R/2$ (strictly greater than $R+\delta$ and strictly less than $2R-\delta$ 2R - δ) (**Figure** 2.6). During this first resynchronisation period, it does not send any resynchronisation message and just waits for the synchronisation messages from the other nodes. At the end of this first resynchronisation period, it considers as active those nodes from which it has received a synchronisation message. It should indeed have received the synchronisation messages from all active nodes, since its first resynchronisation period is long enough to ensure this complete reception.

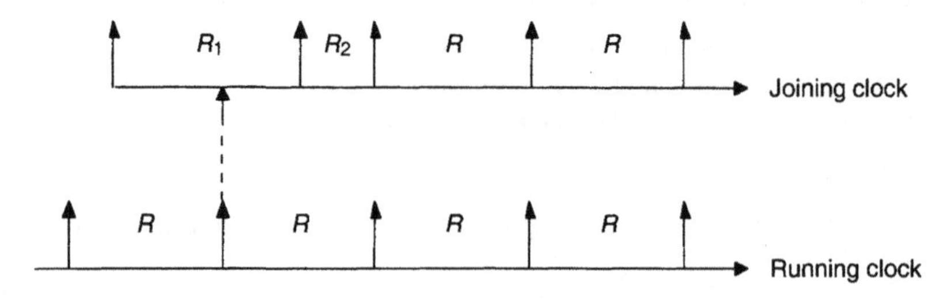

Figure **2.6 — Join Algorithm**

Let T be the arrival date (according to the node's clock) of the latest received synchronisation message[6]. The node sets the duration of its second resynchronisation period, R_2, equal to $R+\Delta$, where $\Delta = T - R_1$. At the end of this second resynchronisation period, the node is synchronised. It starts emitting its own synchronisation message, and enters the nominal synchronisation algorithm.

The Join algorithm synchronises the joining nodes in at most three synchronisation periods of the already synchronised nodes. Thus, the initial synchronisation algorithm could simply consist in successively starting each node every $3R$ seconds.

[6] Notice that the node may have received two synchronisation messages from the same remote node, since the duration of R_1 is strictly greater than R. Moreover, if it is fortuitously started such that it becomes synchronised at the end of R_1, it may have received two synchronisation messages from some remote node and only one from some other remote node. Hence the choice of the latest received synchronisation message for T.

To increase the probability of receiving one and only one synchronisation message from each active node (see footnote 6), the beginning of the second, third and fourth nodes is in fact delayed by a further $R/4$ with respect to the starting of the first node (on which all nodes will normally synchronise). This leads to the following starting scheme:

T0	start node 1
T0 + 3R + R/4	start node 2
T0 + 6R + R/4	start node 3
T0 + 9R + R/4	start node 4

This initial synchronisation algorithm ensures that all nodes are started and synchronised within $10R$ seconds.

2.2 Interactive Consistency

This section describes the ICN exchange protocol [Powell 1997], which basically implements an agreement algorithm with authentication [Lamport *et al.* 1982] and under a hybrid fault model [Lincoln & Rushby 1993]. We recall here that there are two symmetric versions of such algorithms:

- Byzantine Agreement protocol (BA), which allows one transmitter node to send a private value to every other node in such a way that each non-faulty node agrees on the value sent by the transmitter. Moreover, if the transmitter is non-faulty then the agreed value is indeed the private value sent by the transmitter.

- Interactive Consistency protocol (IC), which is a generalisation of Byzantine Agreement where each node sends its private value to every other node.

It has been shown in the literature that, in the presence of m arbitrarily faulty nodes, agreement algorithms using authenticated messages require just $m+2$ nodes[7], instead of $3m+1$ nodes without authentication, and that they must include at least $m+1$ successive rounds of information exchange, sending private values or relaying previously-received values.

Given the full replicated nature of currently-envisaged GUARDS applications, only interactive consistency is really needed and thus only this protocol has been implemented. Should an application require the distribution of a private value from a single channel, then the IC protocol can be used anyway, with the other channels inserting a dummy private value.

2.2.1 The ZA Algorithm

The underlying algorithm of our IC protocol is the ZA algorithm described in [Gong *et al.* 1998]. We describe in this section the (hybrid) fault model, the assumptions used by the algorithm, the algorithm itself, and the properties that it ensures.

---

[7] But note that majority voting on replicated data still requires $2m+1$ nodes.

2.2.1.1 Hybrid Fault Model

When devising practical protocols it becomes natural to distinguish between arbitrarily faulty nodes and nodes affected by less severe kinds of faults (e.g., omission faults). A rigorous formulation of this distinction leads to the so-called hybrid fault model described in [Thambidurai & Park 1988, Lincoln & Rushby 1993]. In this model (which we adopt in the remainder of the document), three types of faults are considered:

- *Manifest faults* produce errors directly detectable by every receiver node through timeout mechanisms, error-coding techniques (possibly including cryptographic checksums) or reasonableness checks.
- *Symmetric faults* deliver the same wrong but locally undetectable value to every receiver node. When authentication is used, any symmetric-faulty relaying node[8] can produce only detectable errors (thanks to the unforgeable signature), and thus symmetric faults reduce to manifest faults.
- *Arbitrary faults* are completely unconstrained, and may deliver (or not deliver in case of omission) any different and/or wrong values to any receiver node.

Under this fault model, when an emitter node (either the initial transmitter node, or a subsequent relaying node) sends a value v to a non-faulty receiver node, the value obtained by the receiver is equal to:

- The value v, if the emitter is non-faulty.
- The distinguished value error, if the emitter is manifest-faulty.
- Some value w (called the value actually sent), if the emitter is symmetric-faulty.
- Any value, if the emitter is arbitrary-faulty.

Note that in the first three cases, all receiver nodes obtain the same value.

2.2.1.2 Assumptions

The ZA algorithm relies on the following assumptions:

A1 Every message that is sent between non-faulty processors is correctly delivered. In particular, link faults are considered indistinguishable[9] from node faults.

A2 The receiver of a message knows who sent it (each node has a private link to the other nodes, cf. Figure 2.4).

---

[8] At the protocol level, a symmetric-faulty transmitter node is indistinguishable from a non-faulty node.

[9] At least, at the protocol level. In a real system, this kind of diagnosis is usually undertaken through autotests. Associating link faults to sender nodes, a fault affecting a single fault containment region can be diagnosed by running the node autotest of the faulty containment region: if the autotest declares the node as non-faulty, then the fault is attributed to the link.

A3 The absence of a message can be detected (since the nodes' clocks are synchronised and the rounds are clock-synchronous).

A4 Messages sent by non-faulty processors cannot be undetectably altered by relaying processors (assumption of authentication).

2.2.1.3 ZA algorithm

The ZA algorithm is derived from the Z algorithm [Thambidurai & Park 1988] by adding the assumption of authentication. In particular, the bug of the Z algorithm pointed out by [Lincoln & Rushby 1993] is precluded by the use of authenticated messages [Gong *et al.* 1998]. The algorithm is defined recursively and proceeds as follows:

***ZA*(0)**

1. The transmitter sends its value to every receiver.
2. Each receiver uses the value received from the transmitter, or uses the default error value in case of a manifest-faulty value.

***ZA*(*r*), *r* > 0**

1. The transmitter signs and sends its value to every receiver.
2. For each p, let v_p be the value receiver p obtains from the transmitter, or error if a manifest-faulty value is received. Each receiver p acts as the transmitter in algorithm $ZA(r\text{-}1)$ to communicate the value v_p to the other n-2 receivers.
3. For each p and q, let v_q be the value receiver p received from receiver q in step 2 of algorithm $ZA(r\text{-}1)$, or error if a manifest-faulty value was received. Each receiver p calculates the majority value among all non-error values v_q received (at most n-1). If no such majority exists, the receiver uses some arbitrary, but functionally determined value.

Note that the number of rounds in algorithm $ZA(r)$ is equal to r+1.

2.2.1.4 Properties

Interactive consistency requires that the following properties hold:

Agreement

If receivers p and q are non-faulty, then they agree on the value ascribed to the transmitter.

Validity

If receiver p is non-faulty, then the value ascribed to the transmitter by p is:

- The value actually sent, if the transmitter is non-faulty or symmetric-faulty,
- The distinguished value error, if the transmitter is manifest-faulty.

Termination

The execution of the protocol terminates in a bounded time.

Under the previous assumptions, [Gong *et al.* 1998] showed that the ZA algorithm satisfies the *Agreement* and *Validity* properties, provided that the following conditions hold[10]:

$$n > a + s + m + 1$$

$$r \geq a$$

with:

n	number of nodes (in our case, we have $n = C$, the number of channels)
a	number of arbitrary faults
s	number of symmetric faults
m	number of manifest faults
r	number of rounds

The *Termination* property is trivially satisfied since, for a given maximum number of faults m, the algorithm executes in $m+1$ clock-synchronous rounds.

2.2.2 The GUARDS Interactive Consistency Protocol

The GUARDS instances have at most four channels. Thus, we consider in the remainder of this chapter architectures with three or four nodes.

The choice of a broadcast network (cf. Figure 2.4) allows us to spread the rounds of the ZA algorithm over successive phases of the protocol, in which a node can send one message to every other node and simultaneously receive three messages (one from each remote node).

For performance reasons, it should be noted that the signatures in the last round can be skipped, because messages of this last round are not relayed. In fact, [Gong *et al.* 1998] claim that is sufficient to use signatures only in the first round. Anyway, in our 2-round implementation of the algorithm, this leads to the same signature policy: sign the messages of the first round and do not sign the messages of the second round. We have taken into account this property in the protocol, but not (for clarity reasons) in the figures.

2.2.2.1 Practical Properties

The protocol implements two rounds of message exchanges between three or four nodes. In conformance with theoretical results (see Section 2.2.1.4), we claim that:

- When $n = 4$, one arbitrary fault (at most) can be tolerated. Of course this implies that one symmetric or one manifest fault can be tolerated as well.
- When $n = 3$, two simultaneous faults can be tolerated, provided that at least one of them is not arbitrary[11]. For example, one arbitrary fault and one manifest fault can be tolerated simultaneously.

---

10 In case n=2, the Agreement property becomes vacuous, but the *Validity* property must still be satisfied.

11 This restriction could be eliminated by using a three-round protocol.

Note that in a practical system, inter-channel fault tolerance by itself is not sufficient: when a channel is faulty, it must be identified (through fault diagnosis) and eliminated (through reconfiguration). However, to simplify the design, we restricted our diagnosis and reconfiguration algorithms (see Chapter 4) to only take into account the case of just one arbitrary fault, both for three and four nodes.

2.2.2.2 Notation

We denote p, q, r, s the indexes of the nodes connected through the ICN, and we describe the protocol from the local node viewpoint. We use the following notation:

V_p

Variable holding the private value of node p.

$V_q(p)$

Variable holding the estimation by node q of private value of node p.

p_broadcast(in msg)

Function allowing node p to broadcast a message msg to every other nodes through the ICN. Each node q, r, s will receive the broadcast message on its private ICN link coming from node p.

msg := p_receive()

Function allowing a node to receive on its private ICN link coming from node p a message msg broadcast (or relayed) by node p. If no message is received during the phase, then msg is set to the default omission message.

v:p := p_encode(in v)

Function allowing node p to encode a value v according to its private key. It returns the encoded value v:p. Note that a node q cannot perform this function since it has no knowledge of p's private encoding key.

v := p_decode(inout msg)

Function allowing a node to decode (by using the public key of node p) a received message msg encoded by node p[12]. If the message is incorrectly encoded or set to the default omission message then the returned value v is set to the default error value and msg is set to the default error message[13], else v is set to the correctly decoded value.

---

[12] The message was either emitted directly by *p*, or relayed by another node.

[13] If the message was already set to the default error message, it is not necessary to re-set it to this default error message. Moreover, this possible modification of the message explains the inout mode of the msg parameter.

v := vote2(in v1, in v2)

Function allowing a node to vote on two decoded values. It is used in the three node configuration. The two-way voting algorithm is given in Figure 2.7.

v := vote3(in v1, in v2, in v3)

Function allowing a node to vote on three decoded values. It is used in the four nodes configuration. The three-way voting algorithm is given in Figure 2.8.

```
if (v1 = v2) then
v := v1;                                -- nominal case: v = v1 = v2
else
        if (v1 = error) then
        v := v2;                        -- detected error on v1
        elsif (v2 = error) then
                v := v1;                -- detected error on v2
        else
                v := error;             -- Byzantine case
        endif;
endif;
```

Figure 2.7 — Two-way Vote

```
if ((v1 = v2) or (v1 = v3)) then
v := v1                                 -- v1 holds the majority value
        elsif (v2 = v3) then
                v := v2                 -- v2 holds the majority value
        else
                v := error              -- Byzantine case
endif
```

Figure 2.8 — Three-way Vote

We use also some intermediate local variables, such as msg1, msg2, ... , msgN for storing the received messages on successive phases or p1, p2, ... , pN for storing the corresponding decoded values.

2.2.2.3 Principle of the Protocol

Let n be the number of nodes. To perform interactive consistency, a first solution would have been to concatenate n Byzantine agreement phases, thus leading to a n+2 phase protocol:

```
phase 1: each node broadcast its private value
phase 2: each node broadcast its estimate of p's value
phase 3: each node broadcast its estimate of q's value
phase 4: each node broadcast its estimate of r's value
phase 5: each node broadcast its estimate of s's value
phase 6: vote values
```

In fact, it is possible to do the job with just n+1 phases, by making use of the symmetry of the problem: each node first sends its private value (round 1) and then successively relays the private values it receives from other nodes in a circular fashion (round 2).

Figure 2.9 shows an example for three nodes where each node sends its private value to every other node. Each box represents the incoming and outgoing links of a given node. Thus, each row of the figure corresponds to a phase of message exchanges (in this example, the three phases correspond to the two rounds of the algorithm), and each column to an emission (noted ↓) and/or a reception (noted ↑) of a given node. The authentication of a message by a node p is denoted by adding the suffix ":p " to the message.

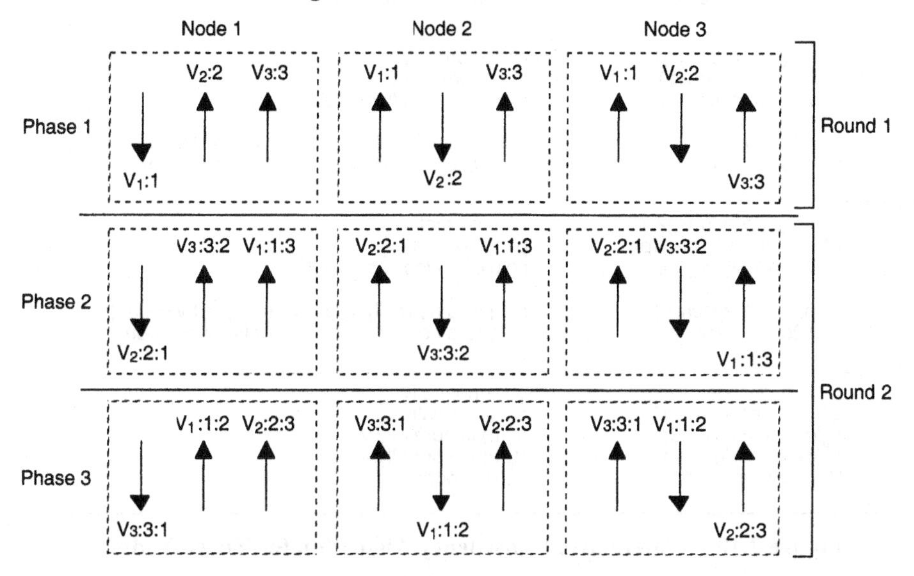

Figure 2.9 — Interactive Consistency Exchange: Example with Three Nodes

In interactive consistency, each node performs exactly the same protocol. The algorithms for three and four nodes are given respectively in Figure 2.10 and Figure 2.11.

2.3 Message Authentication

The GUARDS interactive consistency protocol relies on the assumption of unforgeable signatures. In other words, it is assumed that when a channel *B* receives a message from a channel *A*, it cannot forward an altered version of the message towards a channel *C* without having the channel *C* able to detect that alteration.

In the literature, this assumption is usually substantiated by the use of cryptographic signatures, using a public key cryptosystem such as RSA [Rivest *et al.* 1978]. However, in practice, the implementation of such a scheme can pose serious performance problems. We therefore designed a more efficient scheme, called keyed CRC, which is able to tolerate only accidental faults (intentional faults are not taken into account in the GUARDS architecture).

```
/* Algorithm for node p. The protocols respectively performed by nodes q and r are obtained by a
circular permutation of p, q, and r. */

phase 1
        vp:p := p_encode(vp);
        p_broadcast(vp:p);
//
        msg1q := q_receive();          -- msg1q is normally vq:q
        msg1r := r_receive();          -- msg1r is normally vr:r

phase 2
        q1 := q_decode(msg1q);         -- q1 is normally vq
        r1 := r_decode(msg1r);         -- r1 is normally vr
        p_broadcast(msg1q);            -- no p encoding (last relay)
//
        msg2r := q_receive();          -- msg2r is normally vr:r      (relayed by node q)
        msg1p := r_receive();          -- msg1p is normally vp:p      (relayed by node r)

phase 3
        r2 := r_decode(msg2r);         -- r2 is normally vr
        p1 := p_decode(msg1p);         -- p1 is normally vp
        p_broadcast(msg1r);            -- no p encoding (last relay)
//
        msg2p := q_receive();          -- msg2p is normally vp:p      (relayed by node q)
        msg2q := r_receive();          -- msg2q is normally vq:q      (relayed by node r)

phase 4
        p2 := p_decode(msg2p);         -- p2 is normally vp
        q2 := q_decode(msg2q);         -- q2 is normally vq
        Vp(p) := vote(p1, p2);         -- Vp(p) is normally vp
        Vp(q) := vote(q1, q2);         -- Vp(q) is normally vq
        Vp(r) := vote(r1, r2);         -- Vp(r) is normally vr
```

Figure 2.10 — Interactive Consistency Algorithm for Three Nodes

2.3.1 Algorithm Selection

To substantiate the unforgeable signature assumption, three main schemes are possible (in decreasing induced overhead):

- Cryptographic signatures with public keys (e.g., RSA).
- One-way hash-coding functions (e.g., MD5 [Rivest 1992]).
- Ordinary checksums (e.g., a CRC).

The last two possibilities have to be complemented with private keys in the following way:

- Each pair of channels (i, j) knows privately a common key K_{ij}
- When a source channel i has to broadcast data D, it sends a message M to each destination channel j (in the first round of interactive consistency) equal to D:{f(D,K_{ij}), i<>j}, i.e., it appends to the data D a signature, obtained through a function f (e.g., MD5 or a CRC), such that the signature depends on the transmitted data and the common key K_{ij}.
- In the second round of the IC algorithm, channels just relay the received message (without appending any other signature).

```
/* Algorithm for node p. The protocols respectively performed by nodes q , r and s are obtained by a
circular permutation of p, q, r and s. */

phase 1
        vp:p := p_encode(vp);
        p_broadcast(vp:p);
//
        msg1q := q_receive();           -- msg1q is normally vq:q
        msg1r := r_receive();           -- msg1r is normally vr:r
        msg1s := s_receive();           -- msg1s is normally vs:s

phase 2
        q1 := q_decode(msg1q);          -- q1 is normally vq
        r1 := r_decode(msg1r);          -- r1 is normally vr
        s1 := s_decode(msg1s);          -- s1 is normally vs
        p_broadcast(msg1q);             -- no p encoding (last relay)
//
        msg2r := q_receive();           -- msg2r is normally vr:r     (relayed by node q)
        msg2s := r_receive();           -- msg2s is normally vs:s     (relayed by node r)
        msg1p := s_receive();           -- msg1p is normally vp:p     (relayed by node s)

phase 3
        r2 := r_decode(msg2r);          -- r2 is normally vr
        s2 := s_decode(msg2s);          -- s2 is normally vs
        p1 := p_decode(msg1p);          -- p1 is normally vp
        p_broadcast(msg1r);             -- no p encoding (last relay)
//
        msg3s := q_receive();           -- msg3s is normally vs:s     (relayed by node q)
        msg2p := r_receive();           -- msg2p is normally vp:p     (relayed by node r)
        msg2q := s_receive();           -- msg2q is normally vq:q     (relayed by node s)

phase 4
        q2 := q_decode(msg2q);          -- q2 is normally vq
        p2 := r_decode(msg2p);          -- p2 is normally vp
        s3 := s_decode(msg3s);          -- s3 is normally vs
        p_broadcast(msg1s);             -- no p encoding (last relay)
//
        msg3p := q_receive();           -- msg3p is normally vp:p     (relayed by node q)
        msg3q := r_receive();           -- msg3q is normally vq:q     (relayed by node r)
        msg3r := s_receive();           -- msg3r is normally vr:r     (relayed by node s)

phase 5
        p3 := p_decode(msg3p);          -- p3 is normally vp
        q3 := q_decode(msg3q);          -- q3 is normally vq
        r3 := r_decode(msg3r);          -- r3 is normally vr
        Vp(p) := vote(p1, p2, p3);      -- Vp(p) is normally vp
        Vp(q) := vote(q1, q2, q3);      -- Vp(q) is normally vq
        Vp(r) := vote(r1, r2, r3);      -- Vp(r) is normally vr
        Vp(s) := vote(s1, s2, s3);      -- Vp(s) is normally vs
```

Figure 2.11 — Interactive Consistency Algorithm for Four Nodes

Since in GUARDS only accidental (as opposed to intentional) faults are taken into account, there is no need to use an MD5 signature rather than a plain CRC checksum. Indeed, with the same number b of bits for the checksum as for the signature, one can argue that in both cases, there is a probability in the order of $1/2^b$ for an accidental fault to generate the right checksum for the wrong (falsified) contents. We therefore chose to use a keyed CRC algorithm.

2.3.2 Keyed CRC

Each channel is given a different private key for communication with the each of the other three channels:

Channel 1 <-> Channel 2: key K12
Channel 1 <-> Channel 3: key K13
Channel 1 <-> Channel 4: key K14
Channel 2 <-> Channel 3: key K23
Channel 2 <-> Channel 4: key K24
Channel 3 <-> Channel 4: key K34

Each channel keeps a "key" in read-only memory consisting of the three private keys its needs to communicate with each of the other three channels:

Channel 1: K12; K13; K14
Channel 2: K12; K23; K24
Channel 3: K13; K23; K34
Channel 4: K14; K24; K34

When, in the first round of an interactive consistency protocol, an emitter channel has to broadcast to all other channels a data item D, then each intermediate relayer channel in the IC protocol must be able to:

- Check the integrity of the message (for local use).
- Forward unfalsified copies of the message to the two other receivers.

The first point can be either checked by using a keyed CRC checksum (thus leading to the *totally* keyed CRC algorithm), or by using an underlying low-level checksum provided by some hardwired logic on the 68360 board (thus leading to the *partially* keyed CRC algorithm).

2.3.2.1 Totally Keyed CRC

For $C = 4$, in the first round of an interactive consistency exchange, the emitter (e.g., channel 1) sends to the three other channels an identical message M:

Channel 1 -> Channel 2: M_{12} = M = D:{f(D, K12), f(D, K13), f(D, K14)}
Channel 1 -> Channel 3: M_{13} = M = D:{f(D, K12), f(D, K13), f(D, K14)}
Channel 1 -> Channel 4: M_{14} = M = D:{f(D, K12), f(D, K13), f(D, K14)}

In the second round, the relayer channel (e.g., channel 2) uses the f(D,K12) checksum to check the integrity of the message M, and then forwards M as it is to channels 3 and 4, without adding any further checksum:

Channel 2 -> Channel 3: M_{23} = M = D:{f(D, K12), f(D, K13), f(D, K14)}
Channel 2 -> Channel 4: M_{24} = M = D:{f(D, K12), f(D, K13), f(D, K14)}

Then, channels 3 and 4 are able to check the integrity of the messages M_{23} and M_{24} relayed to them by channel 2, using respectively the checksums f(D, K13) and f(D, K14).

For $C = 3$, only two checksums need to be appended to the first-round message. For $C = 2$, the IC problem is vacuous. However, one checksum is still needed, to locally check on each channel the integrity of the received messages.

2.3.2.2 Partially Keyed CRC

For $C = 4$, in the first round of an interactive consistency exchange, the emitter (e.g., channel 1) sends to the three other channels the following messages:

Channel 1 -> Channel 2: M_{12} = D:{g(D), f(D, K13), f(D, K14)}
Channel 1 -> Channel 3: M_{13} = D:{f(D, K12), g(D), f(D, K14)}
Channel 1 -> Channel 4: M_{14} = D:{f(D, K12), f(D, K13), g(D)}

where g(D) represents the checksum automatically generated by the underlying hardware.

In the second round, the relayer channel (e.g., channel 2) forwards copies of D to channels 3 and 4, along with the corresponding checksum:

Channel 2 -> Channel 3: M_{23} = D:{g(D), f(D, K13)}
Channel 2 -> Channel 4: M_{24} = D:{g(D), f(D, K14)}

Then, channels 3 and 4 are able to check the integrity of the messages M_{23} and M_{24} relayed to them by channel 2, using respectively the checksums f(D, K13) and f(D, K14).

For $C = 3$, only two checksums need to be appended to the first-round message. For $C = 2$, the IC problem is vacuous. However, the checksum g(D) is still needed, to locally check on each channel the integrity of the received messages.

2.3.3 Totally Keyed vs. Partially Keyed scheme

From a design viewpoint, the partially keyed scheme is more complex than the totally keyed scheme.

From a performance viewpoint (in terms of number of checksums to be computed), the partially keyed scheme is better than the totally keyed scheme for Byzantine agreement (where only one channel has a private value to send), and both schemes are equivalent for interactive consistency (where each channel sends its private value).

Since critical applications are usually fully replicated, the protocol that has been implemented is the full IC protocol. So, we prefer the totally keyed scheme over the partially keyed one.

2.3.4 Authentication Coverage

The important point to note is that the relayer channel (channel 2 in the above examples) does not know a priori either K13 or K14. So, in the event of a faulty channel producing a syntactically correct but false message, the probability P_G of that channel being able to "guess" the correct checksum for that message can be estimated as follows (where K is the considered "unknown" key, K13 or K14):

$$P_G = P_{GK} P_K + P_{G\overline{K}} P_{\overline{K}}$$

where:

P_{GK} Pr{correct guess of checksum | successful guess of value of K}
$P_{G\overline{K}}$ Pr{correct guess of checksum | unsuccessful guess of value of K}

P_K Pr{successful guess of value of K}
$P_{\overline{K}}$ Pr{unsuccessful guess of value of K} ($P_{\overline{K}} = 1 - P_K$)

Assuming equiprobable blind guesses, we have:

$$P_K = 2^{-k}$$

$$P_{G\overline{K}} = 2^{-b}$$

where k and b are the number of bits of the key and the checksum.

Furthermore, in the worst case, we have:

$$P_{GK} = 1$$

so the probability P_G is upper-bounded by $2^{-k} + 2^{-b}\left(1 - 2^{-k}\right)$. A lower bound on the coverage ($= 1 - P_G$) of the keyed CRC scheme is thus given by:

$$\text{Coverage} \geq 1 - 2^{-k} - 2^{-b} + 2^{-(k+b)}$$

Note that, when $k = 0$ (i.e., there is no private key K, or, equivalently, the relayer has prior knowledge of K), this expression gives a lower coverage bound equal to 0. This is because we are taking the worst-case viewpoint that an arbitrarily faulty channel with full knowledge of the data and checksum function *will* generate a correct checksum for a falsified message.

Table 2.2 gives some values of this lower bound for various values of k and b. A pragmatic choice for k and b is "as large as possible given the performance constraints".

Table 2.2 — Lower Bound on Coverage of Keyed-CRC Authentication

		k			
		4	8	16	32
b	4	0.87890625000	0.93383789063	0.93748569489	0.93749999978
	8	0.93383789063	0.99220275879	0.99607855082	0.99609374977
	16	0.93748569489	0.99607855082	0.99996948265	0.99998474098
	32	0.93749999978	0.99609374977	0.99998474098	0.99999999953

Chapter 3
Scheduling

One of the key characteristics of the application domains addressed by architecture is the requirement for predictable fault-tolerant real-time behaviour. This is achieved by controlling how the resources required by the applications are allocated and scheduled at run-time. This chapter, therefore, focuses on the GUARDS' scheduling approach.

Section 3.1 provides some background material on the scheduling problem so that the GUARDS approach can be placed in a wider context. Section 3.2 details how applications are scheduled within a channel, and describes the impact of the inter-channel replication approach [Wellings *et al.* 1998]. Section 3.3 introduces the problem of scheduling the interchannel communication network (ICN). Finally, Section 3.4 considers the requirements placed on the operating system needed to support the overall approach.

3.1 Background

Scheduling theory addresses the problem of meeting the specified timing requirements. Since application timing requirements are usually mapped onto process deadlines the issue of meeting deadlines becomes one of process scheduling. A *deadline* denotes the maximum allowed response time of a process. If a task has a *hard* deadline this means that the consequences of a failure to meet its deadline can be catastrophic, i.e., can lead to environmental harm, human injury or significant economic damage. A *scheduler* provides an algorithm or policy for ordering the outstanding processes in the run queue according to some predefined criteria. Scheduling a set of processes consists of sequencing them in such way that the utilisation of resources optimises some scheduling criterion. Real-time scheduling requires an integrated approach to the management of all critical system resources (processor, memory, I/O and communications) in order to achieve an overall coordinated scheduling policy.

Real-time processes can be grouped into two categories: periodic and aperiodic. *Periodic* tasks arrive at regular intervals, while *aperiodic* tasks arrive irregularly and are usually triggered at random intervals by an action external to the system.

D. Powell (ed.), A Generic Fault-Tolerant Architecture for Real-Time Dependable Systems, 51–69.

Although aperiodic processes have timing constraints imposed on them, they cannot have hard deadlines. Consequently, this distinction is usually confined to soft real-time processes while hard real-time processes are divided into periodic and sporadic processes. For *sporadic* tasks, a minimum delay between any two aperiodic events from the same source is needed. The periodic processes typically arise from sensor data or control loops, while the aperiodic processes generally arise from operator actions or asynchronous events. From a scheduler's point of view, a periodic task represents a task with a known start time and deadline, whereas a non-periodic task can arrive at any time, can be started any time after its arrival, and can have arbitrary deadlines.

Most real-time systems execute on top of an operating system. Existing operating systems used in real-time systems almost always provide some combination of FIFO queuing, fixed priority ordering, or round-robin scheduling algorithm. Simple support for management of a hardware real-time clock is provided, with facilities for process scheduling based on the clock.

3.1.1 Scheduling Models

In general, a scheduling model provides two features:

1. A scheduling algorithm for allocating and ordering the system resources.
2. A means (test) of predicting the worst-case behaviour of the applications when the algorithm is applied.

Various scheduling models have been developed over the last few decades. Each has its own advantages and disadvantages. The choice of which model to use will depend on the particular application characteristics. These characteristics form the application's computational model. The *computational model* defines the form of concurrency supported by the application (for example, tasks, threads, asynchronous communication, etc.) and any restriction that must be placed on application programs to facilitate their timing analysis (for example, bounded recursion). In general, applications may conform to a time-triggered, event-triggered or mixed computational model.

- The *time-triggered computational model* is appropriate for an application consisting of a fixed set of periodic tasks.
- The *event-triggered computational model* is appropriate for an application tasks which are solely sporadic in nature.
- The *mixed computational model* consists of periodic and sporadic tasks that communicate asynchronously via shared memory or some form of message passing.

Early scheduling schemes for single processor systems were based on the cyclic executive. The cyclic executive is appropriate for a fixed set of periodic tasks if it is possible to construct a complete schedule that will force all tasks to run at their predefined rate. This means that the schedulability test is done by constructing the cyclic executive ("proof by construction").

For more flexible computational models, it is necessary to move away from cyclic executives to task (or process) based scheduling.

3.1.2 Task-based Scheduling

To describe task-based scheduling schemes, the following assumptions are made:

1. The application consists of a fixed set of tasks.
2. All tasks are periodic, with known periods (i.e., the task is activated or *released* periodically), executed and then becomes dormant until its next release time) or sporadic with a defined minimum inter-arrival time.
3. All tasks have fixed worst-case execution times.
4. All tasks are allocated unique priorities.
5. All tasks are completely independent of each other.

It is assumed that at some point in time all tasks will be released simultaneously. This represents the maximum load on the processor and is known as a *critical instant*. This constraint is fundamental in determining the schedulability of a given task set.

A periodic task can be either runnable or suspended waiting for a timing event. The runnable tasks are executed in the order determined by their priority. The priority of a task is derived from its temporal requirements. With priority-based scheduling, a high-priority task may be released during the execution of a lower-priority one. In a *pre-emptive* scheme, there will be an immediate switch to the higher-priority task. Alternatively with *non-preemption*, the lower priority task will be allowed to complete before the other executes. Between the extremes of preemption, there are alternative strategies that allow a lower priority task to continue to execute for a bounded time. These schemes are known as deferred preemption or cooperative dispatching. In general, pre-emptive schemes enable higher priority tasks to be more reactive, and hence they are preferred.

A common way of scheduling a mix of periodic and sporadic tasks is by using a static priority pre-emptive scheduler. This scheduler will ensure that the highest priority runnable task is executed at runtime.

Table 3.1 gives a standard set of notation for task characteristics.

If a deadline of a task equals its period (i.e., a task must complete before it is next released), there exists a simple optimal priority assignment scheme known as *rate monotonic* priority assignment. Each task is assigned a unique priority based on its period; the shorter the period, the higher the priority. The priorities are represented by integers; the greater the priority, the higher the integer (that is, $T_i < T_j \Rightarrow P_i > P_j$). Liu and Leyland [Liu & Layland 1973] have shown that this assignment is optimal in the sense that if a task set can be scheduled by any fixed priority assignment scheme then it can also be scheduled by the rate monotonic assignment scheme.

Table 3.1 — Standard Notation for Real-time Task Characteristics

Notation	Description
B	Worst-case blocking time for the task
C	Worst-case computation time (WCET) of the task
D	Deadline of the task
J	Release jitter of the task
N	Number of tasks
P	Priority assigned to the task
R	Worst-case response time of the task
T	Minimum time between task releases
U	The utilisation of each task (equal to C/T)
I	The maximum interference of a task

The schedulability test for rate-monotonic scheduling is based upon the utilisation of the task set. If the condition (3.1) below is true, then all N tasks will meet their deadlines (note that summation calculates the total utilisation of the task set).

$$\sum_{i=1}^{N}\left(\frac{C_i}{T_i}\right) \leq N\left(2^{1/N}-1\right) \tag{3.1}$$

However, the rate monotonic priority assignment scheme is not appropriate for tasks with deadlines less than their period. This means that only a set of periodic tasks with the deadlines equal to their periods can be priority ordered in this way. In reality, sporadic (and aperiodic) tasks need to be included and their deadlines are often shorter than their period. Note that for a sporadic task, a period is interpreted as the minimum inter-arrival interval [Audsley *et al.* 1993a]. Sporadic tasks are often used to encapsulate an error handling routine, or to respond to a warning signal, hence they have short deadlines and seldom invocations. Indeed, many periodic tasks will also benefit from having with deadlines less than their period.

If a deadline of a task is less than (or equal) to its period, an optimal priority ordering can be defined with the *deadline monotonic* priority assignment. Here, the fixed priority of a task is inversely proportional to its deadline; the shorter the deadline the higher the priority. If several tasks have the same deadline, they are assigned a priority in an arbitrary manner. Leung and Whitehead [Leung & Whitehead 1982] have shown that deadline monotonic priority assignment is optimal for static priorities.

Schedulability tests developed for the deadline monotonic approach [Joseph & Pandya 1986, Audsley *et al.* 1993b], like the tests for rate monotonic scheduling, are based on a critical instant concept. If all the tasks can meet their deadlines for executions beginning at a critical instant, then they will always meet their deadlines.

If a task set passes the schedulability test, it will meet all deadlines; if it fails the test, it may or may not fail at run-time. This is because tasks may not execute for their WCET, sporadics may not arrive at their minimum inter-arrival times, and there may not be a critical instant.

3.1.3 Response-time Analysis

With response-time analysis the schedulability test is performed in two stages. First, an analytical approach is used to predict the worst-case response time of each task. The values obtained are then compared with the task deadlines. If the worst-case response time of all the tasks are less than their respective deadlines, the task set is schedulable. However, finding the worst-case response time of a task is not a trivial matter [Audsley *et al.* 1993a].

For the highest-priority task, its worst-case response time will equal its own computation time (that is, $R=C$). Other tasks will suffer *interference* from higher priority tasks. So for a task i:

$$R_i = C_i + I_i \tag{3.2}$$

where I_i is the maximum interference that task i can experience in any time interval $[t, t+R_i)$[1]. The maximum interference obviously occurs when all higher priority tasks are released at the same time as task i. Without loss of generality, it can be assumed that all tasks are released at time 0.

A task j (with a higher priority than task i) will be released a number of times (at least one) within the interval $[0, R_i)$. A simple expression for this number of releases is obtained using a ceiling function ($\lceil\,\rceil$) which gives as a result the smallest integer greater than the fractional number on which it acts:

$$Number\_Of\_Releases = \left\lceil \frac{R_i}{T_j} \right\rceil$$

Each release of task j imposes an interference of C_j on task i. Hence:

$$Maximum\_Interference = \left\lceil \frac{R_i}{T_j} \right\rceil C_j$$

Since each task of higher priority is interfering with task i, the total interference I_i is given by:

$$I_i = \sum_{\forall j \in hp(i)} \left\lceil \frac{R_i}{T_j} \right\rceil C_j$$

where $hp(i)$ is the set of tasks with priorities higher than i. Substituting this value back into Equation (3.2) gives [Joseph & Pandya 1986] :

$$R_i = C_i + \sum_{j \in hp(i)} \left\lceil \frac{R_i}{T_j} \right\rceil C_j \tag{3.3}$$

---

1 Note that as discrete time model is used in this analysis, all time intervals must be closed at the beginning (denoted by '[') and open at the end (denoted by ')').

Although the formulation of the interference equation is exact, the actual amount of interference is unknown as R_i is unknown (it is the value being calculated). In general, there will be many values of R_i that form solutions to Equation (3.3). The smallest such value of R_i represents the worst-case response time for the task. It is possible to solve this equation using an iterative technique [Audsley *et al.* 1993a].

The restriction that tasks be independent of one another is not reasonable for a majority of applications. In order to communicate, tasks pass information between themselves. Data communication is usually based upon either shared variables or message passing. *Shared variables* are objects that more than one task has access to; a communication can therefore proceed by each task referencing these variables when appropriate. *Message passing* involves the explicit exchange of data between two tasks. Both types of task communication are supported through use of either shared memory or a physical message-passing network. However, there are differences in the requirements for task synchronisation; if a shared variable is used, a receiver task can read a variable and not know whether it has been written by the sender task, but if a message is used, a receiver cannot obtain a message before that message has been sent. If a task executes an unconditional message receive when no message is available then it will become suspended until the message arrives. Variations in the form of message-based synchronisation arise from the semantics of the send operation: in synchronous communication the sender proceeds only when the message has been received and in asynchronous communication, the sender proceeds immediately, regardless of whether the message is received or not.

If tasks wish to communicate by shared variables, they must protect themselves from pre-emption to avoid shared data becoming inconsistent. It is, therefore, necessary to introduce some form of synchronisation mechanism. A sequence of operations that must be executed indivisibly is called a *critical section*. The synchronisation required to protect a critical section is known as *mutual exclusion*. *Condition synchronisation* is needed when a task wishes to perform an operation that can only be safely performed if another process has itself taken some action or is in some defined state. An example of condition synchronisation is when tasks exchange data via a buffer.

If a higher-priority task is suspended waiting for a lower-priority task to complete then the higher-priority task is *blocked*. In addition, a case of priority inversion occurs when a high-priority task has to wait for a low priority task, and the low priority task is preempted by medium priority task. Consider, as an example, three tasks T_1, T_2 and T_3 with priorities P_1, P_2 and P_3 where $P_1 > P_2 > P_3$. If T_1 is waiting for T_3 to finish some required computation then T_2 will be executed in preference to T_3. Therefore, T_1 will be delayed further because T_2 is executing, even though the priority of T_1 is greater than T_2.

With *priority inheritance*, a task's priority is no longer static; if a higher priority task is suspended waiting for a lower priority task to complete then the priority of the lower priority task is raised to the level of the suspended higher priority task. In general, inheritance of priority is not restricted to a single step.

In a standard priority inheritance protocol [Sha *et al.* 1990], there is bound on the number of times a task can be blocked by lower priority tasks. *Ceiling priority protocols* give even more guarantees as they ensure that:

- A high priority task can be blocked at most once during its execution by lower priority tasks.
- Deadlocks are prevented.
- Transient blocking (i.e., chains of blocks) is prevented.
- Mutual exclusive access to resources is guaranteed.

The *immediate ceiling priority protocol* [Rajkumar *et al.* 1994] defines the following rules:

- Each task has a static default priority assigned (e.g., by the deadline monotonic scheme).
- Each resource has a static ceiling value defined as the maximum priority of the tasks that uses it.
- A task has a dynamic priority that is the maximum of its own static priority and the ceiling values of any resources it has locked.

This protocol provides the mutually exclusive access to the resource on a single processor system.

If B_i is the maximum blocking time that task i can suffer then for simple priority inheritance model, a formula for calculating B can easily be found. Let K be the number of resources (critical sections). The following equation provides an upper bound on B :

$$B_i = \max_{k=1}^{K} usage(k,i)CS(k)$$

where *usage* is 0/1 function : $usage(k,i) = 1$ if resource k is used by at least one task with a priority less than i, and at least one task with a priority greater or equal to i; otherwise it gives the result 0. $CS(k)$ is the maximum computational cost of executing the kth critical section.

Given that a value for B has been obtained, the response time calculation can be modified to take the blocking factor into account. Equation (3.3) becomes :

$$R_i = C_i + B_i + \sum_{j \in hp(i)} \left\lceil \frac{R_i}{T_j} \right\rceil C_j \tag{3.4}$$

To assume that all periodic tasks are released with perfect periodicity and that all sporadic tasks arrive in their minimum inter-arrival interval is not always realistic. The maximum variation in a task's release is termed its *jitter*. This can occur if, for example, the scheduler mechanism takes a bounded time to recognise the arrival of the task. Release jitter is the difference between the earliest and latest release times of the task. The response time analysis represented in Equation (3.4) is not sufficient when tasks experience release jitter. An extra computation time has to be added to the interference [Audsley *et al.* 1993a]:

$$R_i = C_i + B_i + \sum_{j \in hp(i)} \left\lceil \frac{R_i + J_j}{T_j} \right\rceil C_j$$

Until now all system's overheads (e.g., context switching time) have been ignored. The timing overheads incurred by the operating system kernel must be taken into account during the response time analysis. In general, there are two types of kernel overheads: synchronous kernel overheads that occur during the execution of the task (task related kernel overheads), and asynchronous kernel overheads that occur due to the kernel handling asynchronous events which are not directly related to task execution (e.g., interrupt handling). While task-related kernel overheads can be included in the task's worst-case computation time (C), the overheads which are caused by, say, the real-time clock handler, have to be modelled in a specific pseudo-application task (more details can be found in [Burns & Wellings 1996]).

3.1.4 Multiprocessor and Distributed Systems

Schedulability tests and response-time analysis described above are only applicable for the tasks executing on a single processor. As it is known that optimal scheduling for multiprocessor systems is NP-hard [Garey & Johnson 1979], it is necessary to find ways of simplifying the problem. If all periodic and sporadic tasks are statically allocated to different processors than the rate or deadline monotonic algorithms can be used to assign priorities. Furthermore, the interaction between the tasks on different processors can be restricted to common resource usage. If this approach is taken, schedulability of task sets on different processors will depend on a remote blocking caused by competing access to common data objects in a shared memory.

Atomic, non-blocking access to data objects can be provided using algorithms such as those described by Simpson [Simpson 1990]. These algorithms implement single-processor, or multi-processor with shared memory, access to data objects, without any need for mutual exclusion mechanisms. A reading task always reads the most recent version of the data object. A task can read the current value of a data object, modify its value, and write that value back, without needing to lock the object to prevent conflicting access. To determine the most recent version of the data object it is necessary to associate time with the value stored. This time is termed the *timestamp* of the data object.

In the case of scheduling, moving from multiprocessor shared memory systems to distributed systems does not present a significant change. This is due to the method of static task allocation, which is considered more appropriate for distributed systems. To guarantee the deadlines of tasks that communicate across the network, it is necessary to have both bounded access time to the communications medium and bounded message communication delay. One way to ensure this is to use time division multiple access (TDMA) to partition the bandwidth of the channel in the time domain — each node is statically allocated portions of the communication channel in which it can communicate. Alternatively, priority-based or token-passing protocols can be used.

3.2 Scheduling Execution

The GUARDS architecture (see Chapter 1, Figure 1.1) can be considered as a distributed system consisting of 1 to 4 nodes (channels). Each channel is a shared memory multi-processor. Tasks are statically allocated to channels and are replicated across channels for fault-tolerance. To maintain genericity, the architecture is capable of supporting a range of computational models and consequently a range of scheduling models.

Table 3.2 summarises the timing analysis techniques which are used with the chosen computational and scheduling models.

Table 3.2 — Timing Analysis Techniques

		Scheduling model		
Computational Model	Function release	*Cyclic Executive*	*Cooperative*	*Pre-emptive*
Time-triggered	Periodic	Timing analysis by construction	Response time analysis	Rate monotonic analysis
Event-triggered	Sporadic	N/R	Response time analysis	Response time analysis
Mixed	Periodic & sporadic	N/R	Response time analysis	Response time analysis

The choice of scheduling model for a particular application will depend on a variety of factors such as:

- The computational model of the application.
- The performance requirements of the application.
- Constraints placed on the development task by certification authorities (for example, the particular method of testing or the form of safety case required).
- The ease with which the proposed application can be maintained with the chosen scheduling model.

3.2.1 Cyclic Executive Scheduling Model

If the application consists of a fixed set of purely periodic functions, it may be possible to lay out a complete schedule such that the repeated execution of this schedule will cause all functions to run at their correct rate. The cyclic executive is, essentially, a table of addresses where each address points to a procedure encapsulating all or part of the code for a function. The complete table is known as the major cycle; it typically consists of a number of minor cycles each of fixed duration.

3.2.2 Cooperative Scheduling Model

If an application's requirements cannot be met by the cyclic scheduling model, the next simplest model is one based on cooperative scheduling. Here, each application function is represented as a notional task. The tasks are scheduled at run-time by an application-defined scheduler task. Once scheduled, each application task executes until it voluntarily gives control of the processor back to the scheduler task. Hence, the tasks are cooperating in the scheduling of the system.

The main advantages of the cooperative model are that:

- It retains the non-preemptive style of execution.
- It allows periodic functions, which do not fit into convenient major and minor cycles, to be catered for.
- It allows sporadic functions to be executed.

Its disadvantage is that there are inevitably some overheads introduced by the scheduling scheme and it is necessary to undertake some form of timing analysis to determine if all functions meet their deadlines. The actual overheads incurred will depend on the implementation technique used which, in turn, will depend on the difficulty of meeting the real-time requirements of the application.

3.2.3 Pre-emptive Scheduling Model

Although the cooperative model has many advantages over the cyclic executive model, there may be some systems whose timing requirements cannot be met because of the potentially large time between a high priority task becoming runnable and it actually running. Pre-emptive scheduling has the property that if a high priority task becomes runnable and a lower priority task is currently executing, the lower priority task is stopped from executing and the high priority task is executed. The lower priority task executes again when it is the highest priority runnable task. The main disadvantages are that pre-emptive systems require more kernel support and their behaviour is less deterministic (but fully predictable). The pre-emptive model is the most flexible and the one that presents the greatest challenges. It is the one that has been used in the GUARDS prototypes.

A task has a static (base) priority assigned by using, for example, deadline monotonic priority assignment. It may also inherit a higher dynamic priority due to the operation of the immediate priority ceiling protocol. The dispatcher chooses to run the highest dynamic priority runnable task, pre-empting lower priority tasks when necessary. All communication between software tasks is asynchronous. The intra-channel shared memory facilitates this communication model. The use of round-robin scheduling on the VME bus allows all shared memory accesses to be bounded (this is adequate because it is assumed that the number of processors within a channel is small). Furthermore, the use of a non-blocking algorithm, such as that proposed by Simpson [Simpson 1990] helps to avoid the problems associated with remote blocking.

The schedulability analysis is performed on a per processor basis. Tasks are statically allocated to processors (i.e., no dynamic reconfiguration). However, each processor may have more than one mode of operation. There is a common clock across processors within a channel. All communicated data is timestamped so that its freshness can be determined. The schedulability analysis used is based on calculating the response times of tasks and comparing those response times to the tasks' deadlines. The tasks' deadlines are less than or equal to the tasks' periods.

3.2.4 Handling Replication

For an application task to be replicated, it must behave deterministically and each replica task must process the same inputs in the same order. At any point where there is potential for replica divergence, the channels must perform an interactive consistency agreement. Unfortunately, the cost of executing interactive consistency agreement protocols can be significant. There is therefore a need to keep their use to a minimum.

There is clearly a trade-off between the frequency of performing comparisons between channels and the overheads (and scheduling constraints) imposed by such comparisons. In MAFT [Kieckhafer *et al.* 1988, Hugue & Stotts 1991], which supports a cooperative computational model, the scheduling table for each site is replicated for fault tolerance. Each local scheduler, as well as selecting the next task for executing on its site, also mirrors the operations performed by the other schedulers. Every time a task is scheduled or completes, the local scheduler broadcasts its details to all sites. The schedulers then perform an interactive consistency agreement to determine which task has started or completed. The local scheduler can then decide whether the action undertaken by a remote scheduler was in error (that is, the task scheduled was different from the one that a local scheduler thought should be scheduled). Data messages between tasks are also broadcast to all sites. The overheads associated with these protocols are large and MAFT requires a separate custom-built processor (called the operations controller) as well as an application processor in each channel. However, the frequency of comparisons allowed by this approach ensures an early detection of errors.

In our approach, agreement is not necessary on every scheduling decision. Rather, it is important to ensure that all replicated tasks read the same internal data. With this mechanism, a trade-off can be made with fewer agreement communications (and therefore greater efficiency) against early detection of errors. Whichever approach is taken, it is still necessary to perform, interactive consistency or Byzantine agreement, on single-sourced data such as sensor inputs.

In summary, if each replica does not contain any inherently non-deterministic code (for example, use of the Ada select statement), then it is only necessary to:

1. Perform interactive consistency agreement or Byzantine agreement on single-sourced data.
2. Ensure that all replicas receive the same inputs when those inputs are obtained from other replica tasks (replicated inputs).
3. Perform voting on any vital output.

The following subsections indicate how these three requirements are met.

3.2.5 Agreement on Sensor Inputs

Even when replicated sensors are provided, the values read from them must be considered as single-source values and thus potentially different, even in the absence of faults. In order that the replicated tasks in each channel process the same inputs, it is necessary to exchange the values read locally by each channel and to agree on the vector of values that each channel thereby obtains. This is the purpose of interactive consistency agreement (the symmetric equivalent of Byzantine agreement) [Pease *et al.* 1980].

The agreed vector of values is then processed by an application-dependent fault-tolerant algorithm, such as mid-value select, that computes a single value from the agreed vector of possibly erroneous input values. To reduce the complexity of the input selection algorithm, it is important to minimise the error between the redundant input values. However, since the tasks are independently scheduled on each channel, they could read their corresponding sensors at significantly different times. This is similar to the input jitter problem where a task (τ) implementing a control law has to read its input on a regular basis. If jitter is a problem, the solution is to split the task into two tasks (τ^{ip},τ'). τ^{ip} has a release time[2] and a deadline appropriate for the dynamics of the physical quantity being measured by the sensor (to facilitate reaching an agreement on the value of the readings between the replicated sensors). τ' has the original τ's deadline and is executed at an offset from the release time of τ^{ip}. What value this offset should have is discussed in Section 3.2.10.

3.2.6 Identical Internal Replicated Input

Two cases need to be considered when reader and writer tasks share the same data, according to whether or not there is an explicit precedence constraint between the writer and the reader.

When there is such a constraint, then it can be captured by the scheduling. For example, consider the following types of *synchronous* interactions:

S1 Periodic writer – Periodic reader: the writer is given a higher priority than the reader and will produce its data first.[3]

S2 Periodic writer – Sporadic reader: the writer releases the reader.

S3 Sporadic writer – Sporadic reader: the writer releases the reader.

In all these cases, if the writer replicas receive the same input and are deterministic then the reader replicas will always receive the same values. Note that if both a Sporadic writer and a Periodic reader are used then there cannot be an explicit precedence constraint between them but only asynchronous communication (see below).

When tasks share data *asynchronously* (and therefore there is no explicit precedence constraint between the writer and the reader), there are four types of interaction.

---

2 We assume that all I/O is periodic in nature.

3 This is providing the writer does not block and is on the same processor as the reader. If it does block or they are on different processors, the reader must have an offset from the writer equal to at least the response time of the writer.

A1 Periodic writer – Periodic reader: the periods of the two tasks do not have a simple relationship.

A2 Periodic writer – Sporadic reader: the reader is not released by the writer.

A3 Sporadic writer –Sporadic reader: the reader is not released by the writer.

A4 Sporadic writer – Periodic reader: there is no relation between the release of the writer and the reader.

In all of these cases, to ensure each replica reads the same value, more than one copy of that data have to be kept (usually two is enough) and timestamps used [Barrett *et al.* 1995, Poledna 1998]. The essence of this approach is to use off-line schedulability analysis [Audsley *et al.* 1993a] to calculate the worst-case response times of each replicated writer. The maximum of these values is added to the release time of the replicas (taking into account any release jitter) to give a time by which all replicas must have written the data (in the worst case). To allow for clock drift between the processors hosting the replicas, the maximum drift, ε, is also added. This value is used as a timestamp when the data is written. Hence:

$$\text{Timestamp} = \text{release time of writer} + \text{worst-case release jitter} + \text{worst-case response time of writer replicas} + \varepsilon \qquad (3.5)$$

A reader replica simply compares its release time with this data timestamp. If the timestamp is earlier, then the reader can take the data. If the timestamp is later than its release time, then the reader knows that its replicated writer has potentially executed before the other replicated writers. It must therefore take a previous value of the data (the most recent) whose timestamp is earlier than its release time. All reader replicas undertake the same algorithm and consequently get the same value. The scheme has the following properties:

1. All corresponding replicated data items have the same timestamp.
2. All replicated writers will have completed writing their data by the local time indicated by the timestamp.
3. All replicated readers will read the most recent data that can be guaranteed to be read by all of them.

3.2.7 Periodic Writer

If the writer of the data is periodic (cases A1 and A2 above), then expression (3.5) can be used directly. The release time of the writer replicas is common to all replicas and the worst-case response time for the replicas can be calculated by off-line schedulability analysis. Therefore, each writer replica will timestamp its data with the same timestamp and all replicas will have written this data when the local clock at any site is greater that the timestamp.

3.2.8 Sporadic Writer

If the writer is sporadic (cases A3 and A4 above), the release time used depends on how the sporadic was released.

If the writer is released by an interrupt then there is no common release time between replicas that can be used. It is therefore necessary for the replicas to undertake an interactive consistency agreement (they agree to use the latest local release time as the common release time). Each replica, τ, is split into two tasks τ^1 and τ^2. τ^1 is given its release time when it is released by the scheduler; it writes this value to the ICN. τ^2 is offset from τ^1 by a value which is equal to the release jitter of τ^1 plus its worst-case response time plus ε. The release jitter is any skew that might occur between the time the interrupts are generated in each channel (it is assumed that this is known by the designer).

If the sporadic writer is released by a periodic task, then its release time is equal to the release time of the periodic task plus the worst-case response time of the periodic task at the point at which the sporadic task is released.

If the writer is in a chain of sporadic tasks that originates from a periodic task, it uses a release time equal to the release time of the periodic task plus the sum of the worst-case response times of all the tasks in the chain. If the writer is a sporadic task that is in a chain of sporadic tasks triggered by an interrupt, then it uses the same approach but with the agreed release time of the original sporadic task.

3.2.9 Output Voting

Where output voting is required, it is again necessary to transform the replicated task writing to the actuators into two tasks (τ' and τ^{op}): τ' sends the output value across the ICN for voting, and τ^{op} reads the majority vote and sends this to the actuator. The deadline of τ' will determine the earliest point when the ICN manager can perform the voting. The offset and deadline of τ^{op} will determine when the voted result must be available and the amount of potential output jitter. Hence, the two tasks have similar timing characteristics to the tasks used for input agreement (Section 3.2.5). The main difference is that there is a simple majority vote rather than an agreement protocol involving three separate values.

3.2.10 Handling Offsets

The real-time periodic transaction model that has been developed is depicted in Figure 3.1.

A real-time periodic transaction i consists of three tasks τ_i^1, τ_i^2 and τ_i^3. Task τ_i^1 reads a sensor and sends the value to the ICN manager. Task τ_i^2 reads back from the ICN manager the set of values received from all the replicas; it consolidates the values and processes the consolidated reading and eventually produces some data. It sends this data for output consolidation to the ICN manager. Task τ_i^3 reads from the ICN manager the consolidated output value and sends it to the actuator. All communication with the ICN manager is asynchronous via the channel's shared memory.

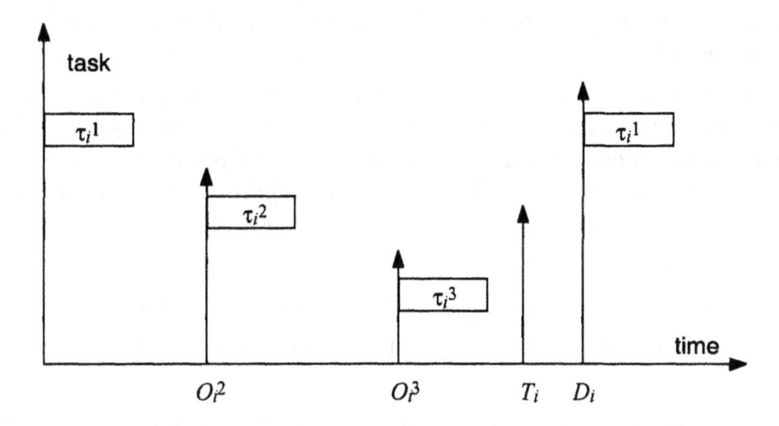

Figure 3.1 — A Real-time Periodic Transaction

This form of real-time transactions is implemented by timing offsets. Analysis of task sets with offsets is N-P complete [Leung & Whitehead 1982] and even sub-optimal solutions are complex [Audsley 1993, Audsley *et al.* 1993c, Tindell 1993]. The approach taken here is based on [Bate & Burns 1997] modified to take into account the fact that the computational times of τ_i^1 and τ_i^3 (respectively C_i^1 and C_i^3) are much less than C_i^2, the computational time of τ_i^2, i.e., $C_i^2 >> max(C_i^1, C_i^3)$. In particular:

1. Priorities are assigned to competing transactions according to transaction deadlines (D) within the following framework. A task τ_i^1 has a greater priority than task τ_j^1, if $D_i < D_j$. All τ^1 tasks have priorities greater than all τ^3 tasks which have priorities greater than τ^2 tasks.
2. Response time analysis is performed [Audsley *et al.* 1993a] on all the τ_j^1 and τ_j^3 tasks and their response times are calculated (assuming that they are all released at the same moment in time; i.e. they form a critical instance). All τ_i^2 tasks are ignored at this stage; they have lower priorities.
3. O_i^2 is assigned to be $T_i/3$ and O_i^3 to be $(2*T_i)/3$ where O_i^2 is the offset of τ_i^2, O_i^3 is the offset of τ_i^3, and T_i is the period of the transaction (see Figure 3.1).
4. A composite task[4] is constructed for τ_i as having a period $T/3$ and a computational time that varies: in its first period it is C_i^2, in its second period C_i^1, in its third period C_i^3 and in its fourth period C_i^2 and so on. This assumes $C_i^2 > C_i^1 > C_i^3$. The order of the C values reflects our assumption about their relative sizes. If this is not the case, the C_i values should be ordered in decreasing sizes. This is so that the composite task generates the maximum

---

[4] A composite task is used for analysis purposes only. It is a task representing the transaction whose real-time properties are chosen so that the interference it causes on lower priority tasks is not optimistic, and yet the interference is less pessimistic than treating the transaction as three tasks that are released simultaneously [Bate & Burns 1997, Bate & Burns 1998].

interference on lower priority tasks, which is essential for the analysis to be non-optimistic.

5. Critical instance response time analysis is used to analyse each τ_i^2 task. However, for each transaction with a higher priority than τ_i^2, when calculating its interference on τ_i^2, its associated composite task is used (i.e., the input and output tasks associated with the higher priority transactions are ignored but the ones from low priority transactions are included). This gives the response times of the τ_i^2 tasks.

Clearly, after the above task, a check must be made to ensure that:

- The response times of the individual tasks are less than the offsets of the next task in the transaction.
- There is enough time before the offset and after the response to transmit data on the ICN network.
- The deadline of the transaction has been met.

If any of these conditions are violated, then it may be possible to modify the offsets of the transaction violating the condition in an attempt to satisfy all the requirements [Bate & Burns 1997].

In the special case where the deadline of the transaction is equal to the period of the transaction then it is possible to optimise the above approach. The first task in the transaction now has the responsibility of writing the result of the voted output (for the previous period) to the actuator and then immediately reading the sensor value. Hence, a transaction now only has two tasks rather than three.

3.3 Scheduling Inter-Channel Communication

The GUARDS ICN has a TDMA access protocol. This is adequate because in the application areas being considered, all input and output is periodic in nature and consequently it is possible to generate a static schedule for the network – even though fixed priority scheduling within each channel is used.

The basic cycle of the ICN is fixed by either the required channel synchronisation precision or by the maximum I/O frequency in a given mode. The cycle is slotted, with each slot being long enough for 1 fixed-sized message transmission (and up to 3 message receptions). As discussed in Section 3.2.10, during system design, each replication application is represented as a real-time transaction. Following the individual schedulability analysis of each channel, the following characteristics are known for each task participating in replicated transactions:

- Period (*T*)
- Response-time (*R*)
- Offset (*O*)
- Deadline (*D*)

From this information, it is possible to build the ICN tables – in the same way as cyclic executive schedules can be constructed (for example, see [Burns *et al.* 1995]). Note that all communication with the ICN manager is asynchronous through the channels' shared memory. Consequently, the ICN manager can take the data any time after the producing task's deadline has expired (see Chapter 7).

Of course, there is a close relationship between the scheduling of the channels and the scheduling of the ICN network. If the off-line tool fails to find an ICN schedule, it is necessary to re-visit the design of the application.

3.4 Operating System Requirements

This section covers the support that commercial off-the-shelf (COTS) operating systems provide for the GUARDS computational and scheduling model. There are several requirements on operating systems kernels that are important:

- Hardware transparency – this includes portability to a variety of hardware platforms, no or very little isolated machine-dependent code, and transparent extension to network operation. Upgrading of the underlying hardware should ideally only require porting of the kernel. Also, a change of kernel should have as little effect as possible on the GUARDS specific hardware components.
- Scalability – a kernel can be scaled down for embedded systems and scaled up for large development workstations.
- Facilities to support distribution (shared memory, network transparency).
- Real-time support – an operating system kernel has to provide a set of basic facilities for real-time applications: clock access, support for different process priority levels, scheduling algorithms, interrupt handling, and pre-emption.
- Inter-process communication through message passing.
- Facilities for spatial and temporal firewalling[5] — the operating system kernel has to provide an environment in which application tasks can execute free from interference from other application tasks (see Chapter 6).

To meet the requirements of the GUARDS architecture and applications, in which instances of this architecture will be used, a decision was made that the operating system kernel has to be compliant to POSIX. POSIX, the Portable Operating System Interface, based on the UNIX operating system, is an evolving set of standards that is being produced by the IEEE and standardised by ANSI and ISO. POSIX is a standard for the source code portability of applications from one system to another. On a system conforming to a particular version of POSIX, the

---

[5] A kernel has a spatial firewalling capability if a memory space with an exclusive rights of access can be allocated to a process. Temporal firewalling capability means that the operating system provides facilities for temporal isolation of tasks and interrupts (e.g., accurately monitoring of task execution time, stopping a task when its allotted execution time has expired, etc).

applications that use the POSIX functions should be able to compile and run with a minimum of modification effort.

The first POSIX standard [IEEE 1003] defines the nucleus of common operating system functions that can be used to create portable programs between different platforms. The operations included are for process creation and deletion, signals, basic I/O and terminal handling. There are two extensions of POSIX that are particularly appropriate for GUARDS. The first addresses those services which are required by real-time applications [IEEE 1003.1b]. The second [IEEE 1003.1c] addresses the notion of concurrent threads executing in a single address space (processes in POSIX execute in separate address spaces).

The POSIX 1003.1b specification provides standardised interface for the following functions:

- Shared memory
- Memory locking and protection
- Asynchronous and synchronous I/O
- Prioritised asynchronous I/O
- Pre-emptive priority scheduling
- Semaphores
- Priority interrupts
- Clocks and timers
- Message queues

POSIX 1003.1b is an interface for a single processor. It does not address explicitly either multiprocessor or distributed systems [Gallmeister 1995]. Some mechanisms provided, e.g., message queues, extend well to a network model while some of the other mechanisms do not extend so well, e.g., shared memory.

POSIX provides suitable mechanisms for spatial firewalls in the form of process abstraction. However, there is no provision for budget timers, which are necessary for the implementation of the temporal firewalls (see Chapter 6).

The operating systems chosen by the GUARDS end-users are QNX and VxWorks. Table 3.3 lists the conformance of these operating systems to the identified POSIX functionalities. Since neither system offers full compliance, a pragmatic approach was chosen: use the POSIX service if it is implemented, otherwise use the equivalent proprietary service.

3.5 Summary

The GUARDS architecture can support a range of real-time computational and scheduling models. We have concentrated primarily on the mixed computational model and the pre-emptive scheduling model. This combination of computational and scheduling models dictates the use of the Response-time Analysis performed on a per processor basis. Communication between applications is asynchronous and facilitated through the intra-channel shared memory. As the number of hosts within a channel is small, all shared-memory accesses can be bounded using round-robin scheduling on the intra-channel VME bus.

Table 3.3 — Profile Functionality for Different Microkernels

POSIX functionality	GUARDS	Microkernels	
	Priority-based pre-emptive	VxWorks	QNX
pthreads	✓	✗ (task)	✗
fork	✓	✗	✓
semaphores	✓	✓	✓
mutexes and condition variables	✓	✗ (mutex semaphore)	✗
message passing	✓	✓	✓
signals	✓	✓	✓
timers	✓	✓	✓
asynchronous I/O	✓	✓	✗
priority scheduling	✓	✓	✓
shared memory objects	✓	✓ (not POSIX)	✓
memory locking	✓	✓	✗
file system	✗	✓	✓

The application tasks that are replicated on different channels must read the same internal data. To achieve this, it is necessary to perform interactive consistency agreement on sensor inputs. However, the overheads associated with executing frequent agreement protocols cannot be tolerated. This, coupled with the need to use fixed priority based scheduling, has led us to use a timestamp approach to achieving replica determinism. However, even this method does not remove the requirement for agreement on sensor input and voting on actuator output. Consequently, we have had to use a real-time periodic transaction model with incorporated offset analysis.

Chapter 4

Error Processing and Fault Treatment

This chapter briefly recalls how fault tolerance is structured in GUARDS and describes in some detail specific procedures error and fault diagnosis and the state restoration.

Fault tolerance is carried out by error processing and fault treatment [Anderson & Lee 1981]. Error processing is aimed at removing errors from the computation state, if possible before failure occurrence, and, in general, involves three primitives [Laprie 1992]:

- *Error detection*, which enables an erroneous state to be identified as such.
- *Error diagnosis*, which enables the assessment of the damage caused by the detected error, or by errors propagated before detection.
- *Error recovery*, where an error-free state is substituted for the erroneous state; this substitution may take on three forms: backward recovery, forward recovery or compensation.

Fault treatment is aimed at preventing faults from being activated again, also involving three primitives [Laprie 1992]:

- *Fault diagnosis*, which consists in determining the cause(s) of error(s), in terms of both location and nature.
- *Fault passivation*, which aims to prevent the fault(s) from being activated again.
- *Reconfiguration*, which consists in modifying the system structure to provide an acceptable, but possibly degraded, service if the system is no longer capable of delivering the same service as before.

In practice, actions aimed at error processing are often interleaved with those related to fault treatment; they do not have to be used in sequence.

The main approach followed in GUARDS for preventing errors from causing failures is to provide enough state redundancy to allow error compensation. Active replication of critical applications (although diversification is not precluded) over the C channels provides this redundancy, which is application transparent and

D. Powell (ed.), A Generic Fault-Tolerant Architecture for Real-Time Dependable Systems, 71–86.

entirely managed by software, above the OS, including voting or comparison. N-modular redundancy is thus adopted to detect the disagreeing channel(s) and (when $C \geq 3$) to mask errors occurring in the voted variables at run-time thus removing errors from the part of the state directly connected to the system output.

State variables (which contain values that are carried over between iterations) are used together with consolidated inputs to compute the output values. However, not all the state variables are consolidated through ICN exchanges and voted (the ICN bandwidth may not support all this traffic and too frequent or heavy consolidations may unacceptably slow down the control process). In addition, no variables in the underlying COTS operating systems can be voted. Therefore, no compensation of the errors in such state and operating system variables ones is performed. This has several implications, among which we recall that: i) even a transient fault can lead to extensive and persistent damage of the internal state of a channel, causing repeated errors in the voted variables, ii) supplementary error processing is necessary in order to preserve the redundancy level of a GUARDS instance (C replicated channels).

The error processing regarding the state of a channel is thus performed resorting to error-detection, error and fault diagnosis and a form of forward recovery called state restoration.

Error-detection events include:

- Discrepancies during inter-channel voting (including timing errors resulting from channel deadline violations).
- ICN bus transmission errors.
- Clock synchronisation errors.
- Inconsistent behaviour during execution of the interactive consistency protocol.
- Incorrect channel status (e.g., wrong frame).

The error diagnosis function collects error reports and is responsible for determining the extent to which the channel's state has been damaged and to decide whether this damage warrants isolation and reintegration. It is based on a filtering mechanism known as α-count [Bondavalli *et al.* 1997a], originally proposed for the discrimination between transient and intermittent-permanent faults.

Later in this chapter, we will describe a family of such α-count mechanisms, and we show how they have been adapted for coping with error filtering to assess the extent of a channel state damage in the GUARDS architecture.

At the inter-channel level, the individual opinions of the channels on each other's state are consolidated in order to find agreement on whether some channel is so extensively damaged as to require state restoration, fault passivation and (possibly) a reconfiguration of the instance.

Once a channel has been diagnosed as requiring passivation, it is isolated (i.e., disconnected from the outside world). Then it is reset and tested in order to determine the nature of the fault. Should it be judged as permanent (the test is not passed), some repair becomes necessary. If the test is passed, the fault is judged as being transient. Note, however, that another α-count could be used to capture intermittent faults which are not caught by the test but hit the channel too frequently.

Whenever a channel passes the test, either because of a transient or after some repair or replacement of components, it must be reintegrated to bring the instance back to its original redundancy level. It must first re-synchronise its clock, then its state, with the pool of active channels. The synchronisation of the re-integrated channel's state is called state restoration (SR) — this is detailed in Section 4.3.

Finally, the last section of the chapter describes the issues related to the implementation of these concepts in the GUARDS prototypes.

4.1 The α-Count Mechanism

The α-count mechanism is a count-and-threshold scheme originally devised "*to consolidate identification of faults, distinguished as transient or permanent/intermittent*" [Bondavalli *et al.* 1997a]. It can be applied on-line in a wide set of application fields, with different dependability requirements. The α-count mechanism tries to balance between two conflicting requirements:

- Signal all components affected by permanent or intermittent faults (referred to in the following as faulty components, or FCs) as quickly as possible. However, gathering information to discriminate between transient and intermittent faults takes time, giving way to a longer fault latency, which in turn imposes requirements on the error processing subsystem.
- Avoid signalling components other than FCs (referred to in the following as healthy components, or HCs); depriving the system of valid resources causes redundancy attrition and may be even worse than relying on a faulty component.

4.1.1 Related Work

The count-and-threshold approach to fault discrimination has long been exploited, for example in several IBM mainframes, and a wide range of techniques have been used, or studied in the literature, with reference to specific systems. Some of them are recalled below, broadly classified into two groups: (A) on-line mechanisms and (B) off-line analysis procedures.

(A): On-line mechanisms

In TMR MODIAC, the architecture proposed in [Mongardi 1993], two consecutive failures experienced by the same hardware component being part of a redundant structure cause the other redundant components to consider it as definitively faulty.

In [Tendolkar & Swann 1982], describing the automatic diagnostics of the IBM 3081, thresholding the number of errors accumulated in a time window is used to trigger the intervention of a maintenance crew. [Spainhower *et al.* 1992] reports how a retry-and-threshold scheme is adopted in the IBM ES/9000 Model 900 to avoid transient errors.

In [Lala & Alger 1988], if a channel of the core FTP can be restarted successfully after error detection (restoring its internal state from other operating channels), then it is brought back into operation. "However, it is assigned a demerit in its dynamic health variable; this variable is used to differentiate between transient and intermittent failures".

In [Agrawal 1988], a list of "suspected" processors is generated during the redundant executions. Several schemes are then suggested for processing this list, including (a) off-line diagnosis of suspected processors, or (b) assigning weights to processors that fail to produce a signature that matches that of the accepted result. In the latter case, those processors whose weight exceeds a certain predetermined threshold are taken off line for diagnosis.

(B): Off-line procedures

In [Tsao & Siewiorek 1983], trend analysis on system error logs is used to attempt to predict future hard failures, by distinguishing intermittent faults (bound to turn into solid failures) from transient faults; the authors conclude that "fault prediction is possible".

In [Lin & Siewiorek 1990], some heuristics, collectively dubbed Dispersion Frame Technique, for fault diagnosis and failure prediction are developed and applied (off-line) to system error logs taken from a large Unix-based file system. The heuristics are based on the inter-arrival patterns of failures (that can be time-varying). Among these, for example, there is the 2-in-1 rule, which signals a warning when the inter-arrival time of two failures is less than 1-hour, and the 4-in-1 rule that fires when four failures occur within a 24-hour period.

In [Iyer *et al.* 1990], a comprehensive and sophisticated methodology for the automatic detection of permanent faults is presented. Error rate is used to build up error groups, i.e., sets of errors occurring in a time interval where a higher than normal rate is observed. The main part of the subsequent analysis exploits the quite detailed information given by the error record structure in the logs generated by the targeted computer systems. Simple probabilistic techniques are recursively used to seek similarities (correlations) among records that may possibly point to common causes (permanent faults) of a possibly large set of errors. The procedure is applied to event logs gathered from IBM 3081 and CYBER machines.

4.1.2 Rationale

The goal in defining the α-count mechanism has been to shape a mechanism simple enough:

- To be implementable as small, low-overhead and low-cost modules, suitable even for embedded real-time systems.
- To allow its behaviour and its effects on the system to be explored using standard analytical means.

An error signalling mechanism is assumed to collect error signals from any available error detection device in the system and to issue error detection results (binary valued signals) to the α-count mechanism on a regular time basis. The judgement on a component's behaviour given by the error signalling mechanism is correct with a probability, or coverage, c. The α-count processes information about erroneous behaviour of each system component, giving less weight to error signals as they get older. The aim is to decide the point in time when keeping a system component on-line is no longer beneficial; if the rate of detected errors, filtered according to some tuning parameters, exceeds a given threshold, a signal is raised to trigger further action (error processing or fault treatment).

A score variable α_i is associated to each not-yet-removed channel i to record information about the errors experienced by that channel. α_i is initially set to 0, and accounts for the L-th judgement as follows:

$$\alpha_i(L) = \begin{cases} \alpha_i(L-1)+1 \text{ if channel } i \text{ is perceived faulty during execution } L \\ K \cdot \alpha_i(L-1) \text{ otherwise, with } K \in [0,1] \end{cases}$$

When $\alpha_i(L)$ becomes greater than or equal to a given threshold α_T, channel i is diagnosed as affected by a permanent/intermittent fault and removed from the system. The effects of applying the mechanism upon system performance and reliability figures depend on the parameters K and α_T. As a first observation, note that $\lceil \alpha_T \rceil$ is the minimum number of consecutive erroneous executions sufficient to consider a component permanently faulty, and K is the ratio by which α_i is decreased after a successful execution. The optimal values to be assigned to these parameters depend on the expected frequency of permanent, intermittent and transient faults and on the probability c of correct judgements of the error signalling mechanism.

It is easy to show that the α–count mechanism is asymptotically able to identify all components affected by permanent or intermittent faults. A (trivially) sufficient condition for α_i to grow greater than or equal to any finite positive integer A used as the value of α_T, is to observe A consecutive erroneous executions. This is also necessary if the parameter K is set to 0. It can also be shown that α–count cannot guarantee not to signal some healthy component as faulty, i.e., a (normally small) probability exists of erroneously identifying a properly working component as faulty.

Many different heuristics of this family can be defined for the accumulation and decay processes providing appropriate filtering actions.

A first direction over which members of the family can be identified is to define the function $\alpha_i(L)$ through slightly different expressions [Rabéjac 1997, Powell *et al.* 1998a]. Instead of a geometric decay we may have a linear one:

$$\alpha_i(L) = \begin{cases} \alpha_i(L-1)+inc \text{ if channel } i \text{ is perceived faulty during execution } L \\ \max\{0, \alpha_i(L-1)-dec\} \text{ otherwise} \end{cases}$$

Moreover, if we consider that several errors can be gathered between any two judgements, and define $e(L)$ as the number of errors collected in such interval, an alternative definition is:

$$\alpha_i(L) = \max\{0, \alpha_i(L-1) - dec + e(L).inc\}$$

In these formulations, a single decision is envisioned about a component, namely: keep it or remove it. Therefore, such a decision directly influences both sides of the trade-off described above. This effect may be attenuated by de-coupling (a) the need to ignore the activity of a possibly faulty component from (b) its actual removal. A new member of the family has been devised for this purpose: a two-threshold α-count scheme [Grandoni *et al.* 1998]. A component *u* is kept in full service as long as its score α_{υ} is less than a first threshold α_L. When the value of α_{υ} grows greater than or equal to α_L, *u* enters a restricted operation mode, where it continues to run application programs, but its results are not delivered to the user or used by the system, other than to continue to feed the α-count. In other words *u* is considered "suspected" and as such its results are not delivered to the user or relied upon by the system. If α_{υ} continues to grow, it eventually crosses a second threshold α_H, with $\alpha_H > \alpha_L$, whereupon the component *u is* considered affected by a permanent fault. If, instead, α_{υ}, after some time, becomes less than α_L, the component is brought back into full service.

This scheme allows the lower threshold to have a low value, thus reducing the diagnosis delay in case of an intermittent/permanent fault. Healthy components being hit by occasional fault bursts are pushed into a non-utilisation period (i.e., until their α-count goes back under the lower threshold), but the probability of irrevocable ousting can be kept as low as desired by means of an appropriately high value for α_H.

Besides being simple enough to be used at run time, these heuristics are also amenable to analysis. Simple analytical models have been derived [Bondavalli *et al.* 1997a, Grandoni *et al.* 1998] which will be described in Chapter 9. They allow a detailed study (using standard tools) of the heuristics' behaviour and the exploration of their effects on the system, in a range of configurations. The performance of the mechanisms has been measured using:

1. The expected value *D*, or the Cumulative Distribution Function (CDF), of the total elapsed time, after the (permanent or intermittent) fault occurrence in a component *u*, for which it is maintained in use and relied upon, because its condition has not yet been recognised.

2. *NU*, the fraction of component life (with respect to its expected life as determined by the expected occurrence of a permanent/intermittent fault) in which a healthy component *u* is not effectively used in the system. *NU* is a measure of the penalty inflicted by the discrimination mechanism on the utilisation of healthy components.

All the heuristics belonging to the α-count family can be easily analysed and provide an appropriate filtering action. So, no restrictions need to be applied on the GUARDS implementation, which can pragmatically choose the easiest heuristic to implement.

4.2 Diagnosis

In the multi-channel architecture adopted in GUARDS, a distributed inter-channel error diagnosis procedure based on α-count is applied. This procedure, filtering the error signals, is responsible for determining the extent of channel's damage. A distributed consensus on the fact that a channel state is corrupted to the extent that it requires further actions to be taken, serves also as fault diagnosis. In fact, it identifies faulty channels on which further fault treatment primitives must be applied.

In the distributed version of α-count, each channel i maintains an α-count representing its opinion of its own health, α_{ii}, and its opinions on the health of the other channels, α_{ij}, $j \neq i$. We assume that the α-counts are updated and processed cyclically (each such cycle is called an α-cycle). The duration, N_{α}, of a α-cycle is a parameter of the mechanism that can be chosen such that $N_{\alpha}.n_1 = N_{frame}$ where n_1 is an integer and N_{frame} the duration of the frame.

The α-count α_{ij} accumulates the opinion of channel i on the extent of corruption of the state of channel j in that it is increased whenever channel i detects an error that it attributes to channel j. Different accumulation weights can be attributed to different error detection events, however we assume for simplicity that all errors are given an equal weight.

Since each channel may have a different perception of the errors created by other channels, the α-counts maintained by each channel must be viewed as private values. Each such value represents that channel's opinion of the extent of state corruption either of itself or of another channel. In order for fault-free channels to have a consistent view of the status of the instance, these opinions must be exchanged through an interactive consistency protocol.

Since the α-count mechanism has the effect of "remembering" detected errors over several cycles, it is important to underline that "fault-free" in this context refers to channels that do not create errors during one execution of an interactive consistency exchange (or, equivalently, during a Byzantine agreement exchange). Consensus can be achieved with C=4 (or with C=3, since signed messages are used) if a single channel is faulty during execution of the protocol, but is provably impossible if more than one channel is faulty. So we must make the following assumption:

> At most one channel may cause errors during the execution of the interactive consistency exchange.

At the end of the last slot of an α-cycle, a local copy α'_{ij} is made of each α-count α_{ij} maintained by channel i. These copies are necessary to "freeze" a value that does not change during the cross-channel consolidation and diagnosis phase. While the latter is being carried out (during the next α-cycle), error events can continue to be accumulated in α_{ij}.

At the beginning of an α-cycle, each channel i thus possesses a vector of α-counts, $\vec{\alpha}'_i$. For C=4, we have:

$$\vec{\alpha}_1' = [\alpha_{11}' \quad \alpha_{12}' \quad \alpha_{13}' \quad \alpha_{14}']$$
$$\vec{\alpha}_2' = [\alpha_{21}' \quad \alpha_{22}' \quad \alpha_{23}' \quad \alpha_{24}']$$
$$\vec{\alpha}_3' = [\alpha_{31}' \quad \alpha_{32}' \quad \alpha_{33}' \quad \alpha_{34}']$$
$$\vec{\alpha}_4' = [\alpha_{41}' \quad \alpha_{42}' \quad \alpha_{43}' \quad \alpha_{44}']$$

After the interactive consistency exchange of the α-count vectors, each fault-free channel k has available a globally consistent α-count matrix, $\alpha_{ii}''(k)$:

$$\alpha''(k) = \begin{bmatrix} \alpha_{11}''(k) & \alpha_{12}''(k) & \alpha_{13}''(k) & \alpha_{14}''(k) \\ \alpha_{21}''(k) & \alpha_{22}''(k) & \alpha_{23}''(k) & \alpha_{24}''(k) \\ \alpha_{31}''(k) & \alpha_{32}''(k) & \alpha_{33}''(k) & \alpha_{34}''(k) \\ \alpha_{41}''(k) & \alpha_{42}''(k) & \alpha_{43}''(k) & \alpha_{44}''(k) \end{bmatrix}$$

such that, for fault-free channels i, j:

$$\alpha''(i) = \alpha''(j) \qquad \text{(agreement)}$$

$$\forall k \in [1, C]: \; \alpha_{jk}''(i) = \alpha_{jk}' \qquad \text{(validity)}$$

A binary accusation matrix $A''(k) = \{A_{ij}''(k)\}$ can thus be generated by comparing the elements of $\alpha''(k)$ with α-count thresholds α_s, for self-accusation, and α_c, for cross-accusation:

$$\forall i, j \in [1, C], \; i \neq j: \qquad A_{ii}''(k) = \begin{cases} F \text{ if } \alpha_{ii}''(k) > \alpha_s \\ C \text{ otherwise} \end{cases}$$

$$A_{ij}''(k) = \begin{cases} F \text{ if } \alpha_{ij}''(k) > \alpha_c \\ C \text{ otherwise} \end{cases}$$

where F and C denote "requiring passivation" and "not requiring passivation".

Note that, instead of consolidating the α-counts, an alternative consists in exchanging binary judgements produced by each channel through a local comparison with α-count thresholds. In this case a consolidated matrix $\mathbf{A}''(k)$ can be built as follows. Each channel i computes its local judgement (at the end of the last slot of each α-cycle):

$$\forall i, j \in [1, C], \; i \neq j: \qquad A_{ii} = \begin{cases} F \text{ if } \alpha_{ii} > \alpha_s \\ C \text{ otherwise} \end{cases}$$

$$A_{ij} = \begin{cases} F \text{ if } \alpha_{ij} > \alpha_c \\ C \text{ otherwise} \end{cases}$$

At the beginning of an α-cycle, each channel i thus possesses a vector of judgements, A_i', i.e., a local test syndrome. For C=4, we have:

$$\vec{A}_1 = \begin{bmatrix} A_{11} & A_{12} & A_{13} & A_{14} \end{bmatrix}$$
$$\vec{A}_2 = \begin{bmatrix} A_{21} & A_{22} & A_{23} & A_{24} \end{bmatrix}$$
$$\vec{A}_3 = \begin{bmatrix} A_{31} & A_{32} & A_{33} & A_{34} \end{bmatrix}$$
$$\vec{A}_4 = \begin{bmatrix} A_{41} & A_{42} & A_{43} & A_{44} \end{bmatrix}$$

After the interactive consistency exchange of the local vectors, each fault-free channel k has available a consolidated matrix, $\mathbf{A}''(k)$:

$$\mathbf{A}''(k) = \begin{bmatrix} A''_{11}(k) & A''_{12}(k) & A''_{13}(k) & A''_{14}(k) \\ A''_{21}(k) & A''_{22}(k) & A''_{23}(k) & A''_{24}(k) \\ A''_{31}(k) & A''_{32}(k) & A''_{33}(k) & A''_{34}(k) \\ A''_{41}(k) & A''_{42}(k) & A''_{43}(k) & A''_{44}(k) \end{bmatrix}$$

such that, for fault-free channels i, j:

$$A''(i) = A''(j) \qquad \text{(agreement)}$$

$$\forall k \in [1, C] : A''_{jk}(i) = A_{jk} \qquad \text{(validity)}$$

Both formulations allow the values to be consolidated at the end of each α-cycle and fault-free channels to have a consistent view of the status of the instance. Thus, both lead to a classic diagnosis problem, in which all channels ("units" in the general literature on diagnosis) can test all other channels, and $\mathbf{A}''(k)$ is the resulting test syndrome, as perceived by all non-faulty channels. This test syndrome can be processed by a fault diagnosis algorithm to generate the final passivation decisions.

The diagnosis problem has been extensively studied in the literature. An ideal diagnosis should be both *correct* and *complete*:

- *Correct* if any channel that is diagnosed as faulty is indeed faulty.
- *Complete* if all faulty channels are diagnosed as faulty.

A diagnosis algorithm is a function that takes as input the test syndrome A and returns a global diagnosis vector **D**, or "pool configuration", whose elements D_i represent the diagnosed state of each channel:

$$D_i = \begin{cases} C \text{ if channel } i \text{ is diagnosed correct} \\ F \text{ otherwise} \end{cases}$$

Note that "correct" (resp. "faulty") in this context means "not requiring passivation" (resp. "requiring passivation").

In the current case the inter-channel tests have imperfect coverage, so a channel requiring passivation is not necessarily accused by all correct channels [Lee & Shin 1990, Blough *et al.* 1992]. The algorithm in the current implementation [Rabéjac 1997] diagnoses a channel as requiring passivation if a majority of channels perceive its state as extensively corrupted, or, of course, if it has this opinion of itself. This algorithm is correct and complete under the assumption that no more than one channel at a time is accused by a fault-free channel. However, due to the memory effect of the α-count mechanism, this assumption can be violated if near-coincident faults occur on different channels. In this situation, there is thus a trade-off between the probability of incorrect diagnosis caused by a long memory effect (high value of K) and the probability of having an incorrect majority vote due to slow elimination of a faulty channel (low value of K). This trade-off is the subject of ongoing research.

Once a channel has been diagnosed as requiring passivation, it is isolated (i.e., disconnected from the outside world). Then it is reset (with the re-initialisation of operating system structures) and a thorough self-test is performed to determine the nature of the fault. If the test is passed the fault is judged as being transient, however another local α-count can be set in order to capture intermittent faults that are not revealed by the test but hit the channel with a too high frequency. If the test is not passed, thus revealing a permanent fault, the channel is switched off and (possibly) undergoes repair.

Whenever a channel passes the test (either because of a transient/intermittent fault, or after repair), it must be reintegrated to bring the instance back to its original redundancy. It must first resynchronise its clock, then its state, with the pool of active channels. This action, a hot restart, can be seen also as a forward recovery action, part of error processing, and it is called here 'state restoration'.

4.3 State Restoration

The problem of state restoration (SR) is so highly application dependent that a general solution is not attainable, and much effort is still devoted to improving the state of the art. A specific solution had thus to be defined for the GUARDS architecture.

SR is a very sensitive operation. It could happen, for example, that one of the active channels, from which the state variables are being copied, is affected by a latent fault that has already corrupted its state but has not yet caused a detectable error. The corrupted state could be transferred to the joining channel, thus setting the stage for a potentially catastrophic failure due to a common-mode error, when the joining channel is put into operation. It is therefore essential to detect errors affecting the state information before activating the joining channel. This can be accomplished through cross-checks during transfer of the state variable data or by some final check executed on the whole state upon completion of the transfer.

In GUARDS, replication is implemented above a COTS OS, which means that the internal state of OS data structures is not accessible, let alone controllable. Furthermore, applications on the rebooted channel must be restarted at a point where the internal state of the restarted OS can be considered equivalent (with respect to the applications) to that of the running channels. This is not simple in a totally pre-emptive environment. It would mean, in general, that all the volatile task state, including stacks, as well as any task descriptors kept by the OS (e.g., open file descriptors), must be copied and/or re-created in the joining channel context.

A reasonable and affordable way to solve the problems above is to exploit the computational model assumed in GUARDS, by awakening the resurrecting replicas exactly at the beginning of their next iteration. Since all tasks start a new iteration at the beginning of an ICN frame, the restart time of a re-aligned channel can be chosen as the beginning of the frame following the completion of state transfer. It has to be assumed that all information carried through successive iterations is stored in the state variables internal to application tasks. SR is then successful if the state transfer operations have correctly brought the joining channel's application context into agreement with the active channels by the beginning of the restart frame.

Another important issue concerns whether the SR tasks can run in parallel with application tasks or not. Obviously, if it were possible to suspend execution of the application tasks, SR would be much simpler. In this case, one-shot SR [Adams 1989] could be performed. This strategy depends on the assumption that the corrupted memory is small enough to be restored at once without degrading critical real-time functions. The implementation of this technique, widely applied in transaction-based systems, is provided among the GUARDS software components, to be used in those specific cases where the assumption is satisfied.

In the general case, however, one-shot SR cannot be used but it is crucial that SR be performed *without* degrading vital real-time functions. It is considered acceptable that the system switches into a special degraded mode, where a reduced number of functionalities representing a minimum level of service is ensured. Hence, the SR task and (possibly a reduced number of) application tasks have to run in parallel. In this scenario, a multi-step approach has to be followed. At each stage, only a fraction of the state information is exchanged, as allowed by the spare ICN throughput. Several stages are thus needed to refresh the entire state. Such multi-step SR is complicated by the fact that application tasks may change values of state variables already restored in the joining channel before SR itself has terminated. Two categories of multi-step SR, referred to as *Running SR* and *Recursive SR* have been identified for GUARDS, described in [Bondavalli *et al.* 1997c, Bondavalli et al. 1998c]. Previous work giving hardware-assisted versions of each can be found in [Arlat *et al.* 1978, Sims 1996].

4.3.1 Running SR

The basic behaviour of Running SR, assuming only one active channel apart the joining one, is the following. Let the channel context be arranged in a single (logical) memory block managed by a "context object". When a channel needs to be re-aligned, the system enters an "*SR mode*". The active channel enters a "*put state*" mode while the joining channel enters a "*get state*" mode. All updates to state variables by the active applications are propagated through the ICN and are thus performed also in the joining channel's memory space.

Two types of tasks are activated: a *Sweeper* task on the active channel and a *Catcher* task on the joining channel. Moreover, the code of the application tasks needs to be extended (with the definition of an extended_assignment procedure) to allow for the parallel update of state variables both on the local channel and on the joining channel (via the ICN). In the frame/cycle/slot timing structure of the ICN, the Sweeper makes use of asynchronous slots in the ICN schedule (i.e., slots left available by cyclic critical applications) to exchange context data. The Sweeper uses

CPU time not used by vital applications, while the Catcher, running on the joining channel, has most of the CPU time available, as the application tasks are waiting to be restarted. The description of the basic behaviour of the *Running SR* scheme in this simple 2-channel GUARDS instance is given on Figure 4.1 to Figure 4.3.

```
/* context_morsel[]:        array of records containing the portion of context read by the Sweeper
                            introduced to optimise the use of the ICN by sending "packets" fitting
                            into the ICN slot; generic record structure: <state variable
                            identifier>,<state variable contents>;                          */
/* end_of_context():        Boolean function: returns TRUE when context_morsel[] contains an
                            (end of context) mark                                            */
{
        do      {
                context_morsel[]=get_next_context_morsel();
                send(context_morsel, Catcher);
                };
        while (! end_of_context(context_morsel[]));
}
```

Figure 4.1 — Algorithm of the *Sweeper* Task

```
/* Substitute each assignment instruction directed to a state variable with the following procedure:
                                                                                              */

extended_assignment (state_variable, new_value);
{
        state_variable=new_value;
        send(state_variable, Catcher);
}
```

Figure 4.2 — Extension to the Code of Application Tasks

Switching from normal computation to the "SR mode" occurs at the beginning of a frame, with a corresponding change in task scheduling, and SR completion always occurs at the end of a frame with normal application scheduling activated at the following one.

The tasks involved in the SR process run asynchronously on two separate physical machines communicating through an asynchronous channel. The problems of their coordination and synchronisation can be solved as follows:

- The active applications and the Sweeper task are assigned deadlines in such a way as to ensure that they are scheduled in the same order as the ICN cycles they use to transfer context information.
- The Catcher task updates the state variables in the same order that they come from the ICN.
- The procedure extended_assignment is executed atomically with respect to the other application programs running on the channel. This ensures that shared variables are accessed consistently, even during the state restoration operations.

```
/* context_morsel[]:        portion of context obtained from the Sweeper for each exchange of data
                            between channels                                                     */
{
    do    {
                for each task_j in {Active_Tasks}
                do      {
                        receive(state_variable, task_j )
                        update_context(state_variable);
                        };
                receive(context_morsel[], Sweeper);
                if (! end_of_context(context_morsel[]))
                update_context(context_morsel[]);
          };
    while ( ! end_of_context(context_morsel[]));
}
```

Figure 4.3 — Algorithm of the *Catcher* Task

In this simple two-channel case, there can be no inter-channel protection from errors affecting the active channel before completion of the SR procedure. If the state of the active channel is corrupted by an error that remains latent despite intra-channel self-checking, this error will be copied across to the joining channel. Note that this reliance on intra-channel self-checking is consistent with the very notion of state restoration in a two-channel architecture since diagnosis of which channel requires state restoration also depends on these intra-channel self-checking mechanisms.

Consider now a more general GUARDS instance, where two or more active channels cooperate to align a joining channel. To take advantage of the parallel links of the ICN, the whole block of memory storing the channel context is split into *C*-1 sub-blocks of similar size, each managed by an active channel. Each active channel *i* propagates to the joining channel the updates performed by running applications to state variables belonging to block *i*, and runs a Sweeper task (on spare CPU time slots) that conveys the *i*-th block of the context. In the joining channel all the information is received and processed by a Catcher task that has most of the CPU time available, as application tasks are not running.

This organisation is too simplistic in that it does not exploit the active channels multiplicity for detecting latent inconsistencies in the replicated variables. The redundancy of the active channels must be exploited during the SR protocol to ensure that the state be correctly restored. Two possible approaches are:

- **Detection of SR errors by comparison**: Each Sweeper sends the state values not only to the Catcher on the joining channel, but also to the other (*C*-2) active channels, so that the latter can compare the received value with their corresponding local data. To cope with inconsistent (asymmetric) failures, a Byzantine agreement must be used to ensure consistency of the data sent to the joining channel and to the other active channels. In addition to the necessary modifications to the Sweeper task's code to restrict its actions to a subset of the entire state memory block, and to the operations for the Byzantine agreement such as the introduction of a R_broadcast() primitive, a new task, called *Checker*, reported in Figure 4.4, has to be activated on each active channel, to compare the data exchanged with the corresponding local value. An inconsistent (asymmetric) failure during the protocol execution is detectable by all channels (active or joining). A consistent (symmetric) failure can only be detected by the other active channel(s). The main drawback of this approach is that the overheads of Byzantine agreement may cancel the performance gains obtained by partitioning the memory block to be transferred.

- **Detection of SR errors by signatures**: Every Sweeper and the Catcher compute a signature of the *whole* local channel state after the state data has been transferred to the joining channel. The signatures are then exchanged through the interactive consistency protocol. For the SR to be considered successful, all signatures in the consistency vector have to match. A disagreeing signature originated by a Sweeper implies that the hosting active channel has failed. A mismatch between the signature computed by the Catcher and those of the Sweepers points to a failure in the joining channel or to a transient failure in one or more active channels during the state transfer (typically, in the associated ICN medium). Should a disagreement occur, the SR procedure has to be repeated. The implementation of this solution requires additional code in the Sweepers and in the Catcher to compute the signatures, to call the interactive consistency procedure, and to inspect the resulting consistency vector. If performance allows, the state signature comparison among active channels could also be used systematically, in every frame, to improve error detection latency. This signature-based scheme appears to be more efficient than the previous one since there is just a small computational and communication overhead, but the error coverage is strictly related to the adopted signature scheme. This is the solution implemented as a GUARDS software component.

```
/* N_Sweepers          constant indicating the number of Sweeper tasks, one on each active
                       channel                                                          */
/* Sweeper_stopped     variable to count how many Sweepers have been stopped            */
/* compare()           procedure that compares the context values received with the
                       corresponding local values. Appropriate signalling is raised upon
                       detecting a discrepancy.                                         */

{
    Sweeper_Stopped=0;
    do  {
            for each channelj in {Other Active Channels}
            do  {
                for each taskk on channelj
                do  {
                        receive(state_variable, taskk )
                        compare(state_variable, local_state_variable)
                    };
                receive(context_morsel[], Sweeper);
                compare(context_morsel[], local_context_morsel[])
                if (end_of_context(context_morsel[]))
                        Sweeper_stopped++;
                };
        }
    while ( (Sweeper_stopped < N_Sweepers -1));
}
```

Figure 4.4 — Algorithm of the *Checker* Task

Since a deterministic, finite time is required to copy the memory block and any update to already copied state variables is carried out directly, the whole (parallel) state restoration is performed in a deterministic, finite time. Note that, during SR, the ICN has to support: i) the normal traffic generated by the vital (i.e., non-stoppable) applications, ii) the extra traffic due to state variable updates, and iii) the traffic generated by the Sweeper task.

4.3.2 Recursive SR

The basic behaviour of Recursive SR, assuming only one active channel apart the joining one, is the following. Consider now the channel context arranged as (or mapped into) an array. Each element holds a state data item (a state variable or a vector of state variables) and a *binary tag*. The value for all the tags is set to *true* when SR starts. During SR, any Write operation on a state data item sets the accompanying tag to *true*. A special tRead operation must be provided, which gets the data item and resets the tag to *false*. Write and tRead are both atomic operations. The context transfer operation is performed sequentially by a Sweeper task. Upon activation, the Sweeper starts reading the context data table, using tRead, and sends the data to the joining channel through the ICN, while the regular applications continue to use normal Read operations. When the Sweeper reaches the end of the table, it starts again from the beginning, looking for any tags that have been set back to *true* by application tasks. If so, the corresponding state variables are re-sent, and so on.

A termination condition must be defined since, unlike *Running SR*, it is not possible to determine in advance the time required for completion. The definition of such a condition, however, does not imply that the condition will necessarily occur: termination is inherently non-deterministic and depends on the detailed timing properties of applications. Of course, it must be ensured that there are no further updates to state variables by application tasks from the moment the condition is satisfied until the end of SR (i.e., the end of the frame during which the condition becomes true). This could be done by suspending applications until the end of the frame. However, this is unacceptable when vital activities have to be kept running. Hence, a careful scheduling of the application tasks and of the Sweeper is necessary.

One possible termination condition is to detect when the number of updated variables has become small enough for them to be transferred in a single non-interruptible operation (depending on the maximum duration for which applications can be suspended). This termination condition, referred to as the "last shot" transfer condition, can be implemented with the aid of a counter variable associated with the array holding the channel context. This counter indicates the current number of elements in the array (that is, state variables) whose associated tag is *true*. At the beginning, when the SR phase starts, the counter is set to the total number of state variables. The counter is decremented by every tRead operation and is incremented by any Write operation that flips the tag from *false* to *true*. A call to termination_condition() checks the counter, which represents the number of state variables still to be transferred: if the transfer can be performed as a "last shot", the function returns *true*.

Recursive SR can be adapted to a more general GUARDS instance, where two or more active channels co-operate to align a joining channel. A few schemes based on sharing the SR activities among the active channels have been identified which induce a higher complexity in defining and verifying the termination condition.

A fully-motivated choice between the two approaches (and their variations) would require a more detailed analysis. In the implementation of the GUARDS prototypes, Running SR has been preferred due mainly to the fact that termination of Recursive SR is inherently non-deterministic and depends on the detailed timing properties of applications. More details on Running SR, and a full description of Recursive SR, with variations and optimisations are given in [Bondavalli *et al.* 1997c, Bondavalli et al. 1998c].

Chapter 5

Output Consolidation

Output consolidation is concerned with the mapping of multiple redundant outputs into a single physical effect on the controlled process. The diversity of the targeted applications and their various dependability objectives leads to different instantiations of the generic architecture with different consolidation strategies for safety-critical and safety-related outputs. The generic output consolidation mechanism presented in this chapter proposes, for safety-related outputs, a cost-effective complement to the application-specific voters necessary for safety-critical outputs. Since a single instance of the architecture may host applications with different criticality levels, it may in fact have both a generic output consolidation mechanism for safety-related outputs and application-specific voters for safety-critical outputs.

A further aspect of output consolidation concerns the disabling of the outputs of faulty channels to prevent voting discrepancies in the event of successive failures of other channels. The designer of a particular instance of the architecture can choose among three channel passivation mechanisms, according to the dependability requirements of the corresponding application:

- *Software-based channel isolation*: an isolation command can be executed by a channel that diagnoses itself as faulty in order to isolate itself from the outside world. This is the simplest passivation mechanism, but it is also the least dependable since it relies on partially correct operation of the channel that is to be disabled. This approach is not detailed further.
- *Hardware-based channel isolation*: each channel contains specific hardware that implements a distributed vote on the pool configuration computed by the software-implemented diagnosis algorithm (cf. Chapter 4) and provides physical isolation of a faulty channel. This mechanism also allows automatic re-insertion of a channel that was off-line, but is then judged to be non-faulty (e.g., after a successful self-test).
- *Hardware switch-off by fail-safe exclusion logic*: as in the previous case, each channel contains specific hardware that implements a distributed vote on the pool configuration. In this case, a channel that is diagnosed as faulty is irreversibly powered off until a manual re-arming operation is carried out.

D. Powell (ed.), A Generic Fault-Tolerant Architecture for Real-Time Dependable Systems, 87–98.

The irreversibility of the isolation and the fail-safe implementation of the specific hardware make this approach the most appropriate for the very critical applications.

The remainder of this chapter is divided into three sections. Section 5.1 discusses the principles of output consolidation and describes the generic output consolidation mechanism that has been designed to provide a cost-effective solution for safety-related outputs. This mechanism is implemented as part of a VME-compatible Redundancy Management Board (RMB), which also provides hardware-based (reversible) channel isolation. This RMB is described in Section 5.2. Finally, Section 5.3 describes the fail-safe exclusion logic that can be used to implement (irreversible) hardware switch-off.

5.1 Consolidation Mechanisms

The first paragraph recalls the principles of the voters used for safety-critical outputs. The following paragraphs are devoted to the safety-related outputs, which are the main focus of the generic output consolidation mechanism.

5.1.1 Safety-Critical Outputs

Safety-critical outputs are commonly consolidated by means of specific voters that are strongly application-dependent in terms of: (i) technology (dynamic logic, fail-safe relays...), (ii) electrical level of the output signals, and (iii) certification requirements for each targeted domain. The voter technology is carefully chosen to facilitate the safety demonstration. Most solutions are based on discrete chips coupled with a Failure Mode Effect and Criticality Analysis (FMECA) analysis. Safety is based on some invariant physical property like: (i) the saturation of a magnetic core for dynamic logic voters, or (ii) earth gravitational force for fail-safe relays in the railway domain.

Single voters such as these are fail-safe but represent a single point of failure with respect to availability. Fail-operational actuation can be achieved by means of multiple actuators, each with a fail-safe voter, as shown in Figure 5.1 for a triple modular redundant (TMR) system.

5.1.2 Safety-Related Outputs

Safety-related outputs, by definition, have less stringent safety requirements than safety-critical outputs. For such outputs, application-specific voters may not be the most cost-effective solution. This is especially the case when multiple output configurations must be covered by a single generic architecture. This is particularly true for controlling a process that requires: (i) a large number of outputs (several tens), (ii) a mix of output signals (digital, analog), (iii) many switches or (iv) a high switching frequency.

The output consolidation mechanism described now is designed to provide a global solution for various applications. The analysis of the requirements from the nuclear, railway and space domains leads to a typical output configuration of 32 digital and 8 analog outputs for instances including 2, 3 or 4 channels.

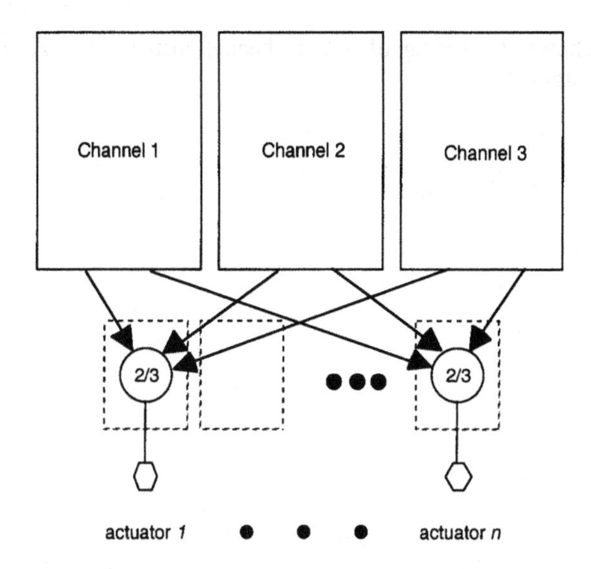

Figure 5.1 — TMR Instance with a 2/3 Voter per Actuator

The main design constraints for the generic output consolidation mechanism are:

- To provide dependable outputs while using COTS I/O boards and minimum specific hardware.
- To be compatible with the different instantiations of the generic architecture.

As a consequence of these constraints, the generic output consolidation mechanism is based on a *read-back* loop instead of specific voters. This technology limits the dependence on the application to the choice of the COTS I/O boards by the end user.

The implementation of the read-back loop requires some external wiring to acquire the actual state of each output. For a C channel instance, each output is acquired by one read-back locally to the channel and by C-1 remote channel read-backs.

An *Output_Manager* software module executed by the host processor(s) of each channel periodically compares the expected output computed by the application with the actual output given by the read-back. The result of these comparisons are error syndromes to be voted at the inter-channel level. This vote is carried out by a specific *Redundancy Management Board* (RMB), which is described in the next section.

When the majority of the channels detects at least one output of a channel in an erroneous state, a corrective action is undertaken. This corrective action puts the faulty channel in the safe state by disconnecting its outputs from the process. To limit the dependence on the application, fail-safe disconnection is implemented outside the boards by means of relays in series with the output signals. As the output consolidation mechanism is fully distributed over the channels, at the system level there is no single point of failure with respect to availability.

Figure 5.2 shows the wiring of a two-channel instance using the read-back principle for *n* outputs.

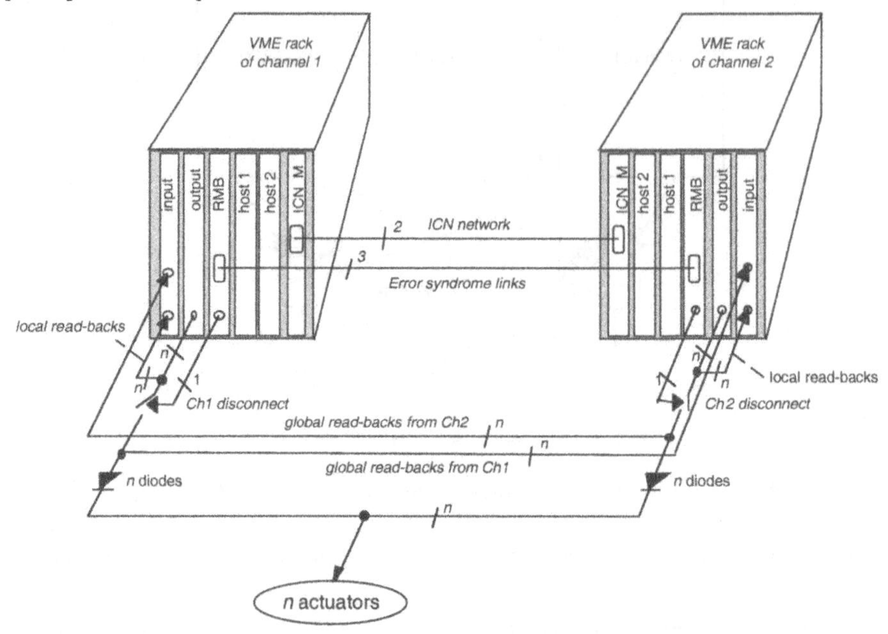

Figure 5.2 — Generic Output Consolidation for a 2-Channel Instance

It is important to note that, by its very principle, the output consolidation mechanism can only react to an error when the output has already been affected: this leads to a transient erroneous state on the actuator command (i.e., the command is not glitch-free). It is assumed that this transient is short enough (a few milliseconds) to be filtered by the actuator, as would be any other electrical perturbation. Even though the response time of the mechanism is short, the transient state will still exist. If this transient is not acceptable for a particular actuator, then a classic voter must be used.

5.1.3 Experimental Results

The described generic output consolidation mechanism is used in the nuclear propulsion prototype (cf. Chapter 10) which is used in a safety-related system with a channel redundancy of two.

This output consolidation approach requires a large proportion of CPU time and is limited by the real-time performance of the operating system. On the prototype, the execution time of the implemented *Output_Manager* service is 1.1 ms. Most of this time is due to waiting loops required to comply with the hardware timing of the employed analog I/O boards. The execution time can be reduced to 30 µs for the consolidation of 16 digital outputs. A similar performance could be obtained with the use of intelligent analog boards providing autoscan functions.

5.2 Redundancy Management Board

The Redundancy Management Board (RMB) is implemented as a VME board for easy integration in existing racks (no external hardware). Each channel has one RMB. The RMB interfaces are: (a) the VME bus, for the commands from the local channel, and (b) dedicated hardware links for communication with the RMBs of the remote channels (cf. Figure 5.2).

The set of RMBs votes the commands received and performs corrective actions (e.g., isolation of a faulty channel) in a consistent way among the channels. The RMB can be strap-configured to implement one of four voting algorithms:

- Two-out-of-four (2/4)
- Two-out-of-three (2/3)
- Two-out-of-two (2/2) without reconfiguration
- Two-out-of-two (2/2) with reconfiguration to simplex (1/1)

The set of RMBs receives the current pool configuration from the pool manager, which implements the diagnosis algorithm described in Chapter 4. This pool configuration is updated after error filtering by the α-count algorithm. The RMB also receives commands from the *Output_Manager* service described in Section 5.1.2.

5.2.1 Inter-Channel Voting

The output consolidation scheme is based on inter-channel voting. Each vote is "the local opinion of each channel about the others" (we use here the term "*disconnection request*" to denote the indication of a discrepancy). The result of these votes gives "the opinion of the majority of the channels" about each channel (we use the terms *disconnection* and *connection* to denote the majority opinion about a given channel).

Taking a TMR instance as an example, the classic approach consists in using as many two-out-of-three voters as actuator commands (cf. Figure 5.1). To ensure the same behaviour while voting on the local opinion of the channels, the number of voters is reduced to three and the voting algorithm is slightly different, as outlined below (and illustrated on Figure 5.3 for the TMR case):

1. The voter is fully distributed across the channels. Each voter computes *connection/disconnection* for the channel where it is located.
2. A key point of the design is to ensure the reciprocity of the *disconnection request*, i.e., when a channel (say channel *i*) disagrees with another channel's result (say channel *j*), channel *i* sends a *disconnection request* to channel *j*. The design ensures that channel *i* also receives a *disconnection request* from channel *j*. This property is necessary to ensure correct behaviour in a two-out-of-two configuration (i.e., the nominal configuration for a two-channel instance or a degraded configuration for a three- or four-channel instance). The necessity of the property becomes apparent when the second fault occurs on Figure 5.3.

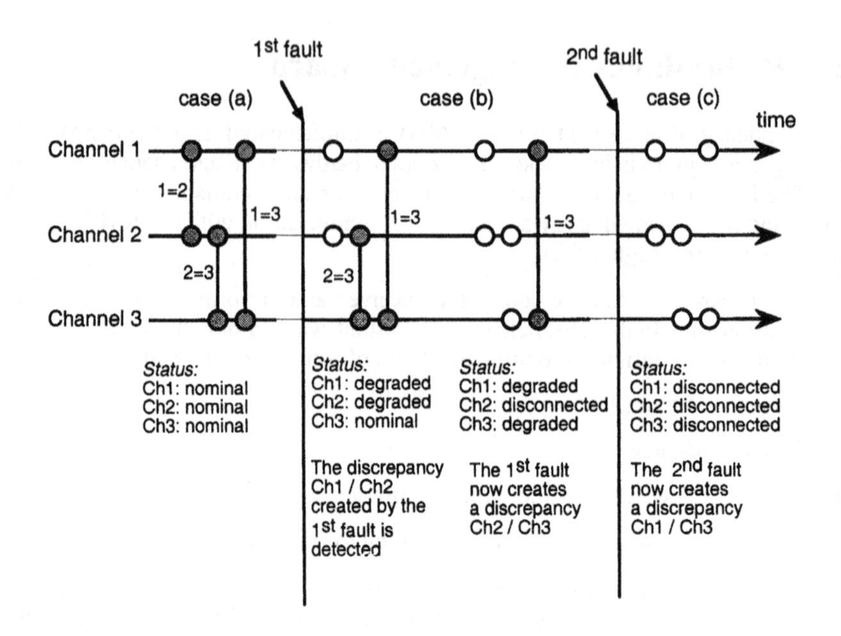

Notation:

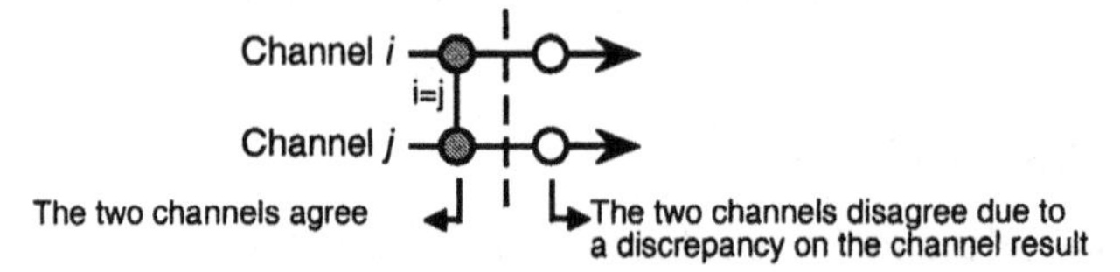

a) nominal, b) degradation after a 1st fault and c) after 2nd fault

Figure 5.3 — Different Configurations of a TMR Instance

3. The design ensures the disconnection of the faulty channel when it one receives more than one *disconnection request.* Receiving only one *disconnection request* warns the channel of the detection of a "non-confirmed discrepancy" involving itself and the sender. Receiving two *disconnection requests* means, for a TMR instance, the detection of a "confirmed discrepancy"involving itself and two different senders. In other words, two channels ask for the elimination of the third. This property ensures the degradation of the instance to (i) 3 channels (first fault of a 4-channel instance) and to (ii) 2 channels (second fault of a 4-channel instance or first fault of a 3-channel instance).

4. The design ensures that a channel is reconnected when it and the remote channel agree with each other. This property allows an instance to successively grow from 2 channels to 3 channels, and then to 4 channels.

5. The design ensures that an instance can be started in either a nominal configuration (e.g., 3 channels healthy for a TMR) or a degraded configuration (e.g., 2 channels healthy for a TMR).

6. The design ensures that an unpowered channel (faulty or not) sends a *disconnection request* to the other channels.

5.2.2 Detailed Design

The RMB is hardware-implemented as a VME board, using field-programmable gate arrays. It has three different subparts (see Figure 5.4):

1. *The VME interface with the host CPUs*

   The VME interface is in the same fault containment domain as the host CPUs and thus has a similar objective in terms of self-checking coverage.

2. *The voter itself*

   The voter takes decisions at the instance level and must achieve the same level of self-checking coverage as that of the set of redundant channels. This requires structural redundancy within each board: the voter is made up of two independent "data paths", which are cross-checked by a comparator in each path.

3. *The data path comparators*

   The data path comparators require the same self-checking coverage as the voter. However, the detection of a discrepancy between the two data paths is characteristic of an error inside the RMB. The RMB goes into a non-recoverable safe state called emergency mode as the fail-safe property is no longer guaranteed for a further fault. The current RMB is a functional prototype that only implements a single data path.

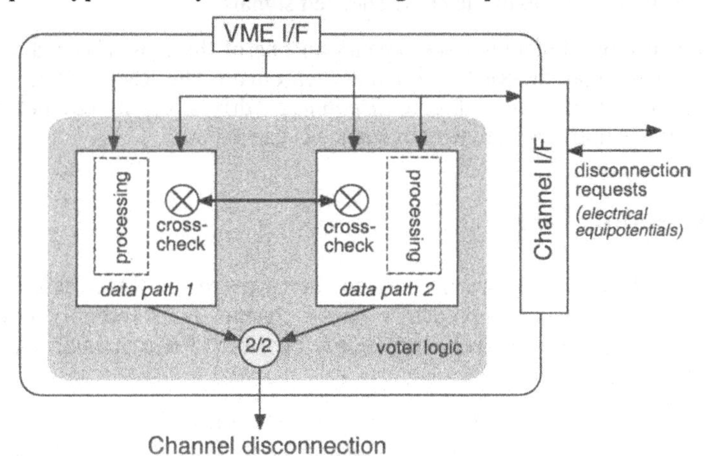

Figure 5.4 — Outline of RMB Board

5.2.2.1 Processing Performed by the VME Interface

The VME interface dispatches the commands received from the VME bus to each data path. The commands are encoded by the emitter (host CPU) to ensure error detection. An identification of the emitter is also included.

5.2.2.2 Processing Performed by each Data Path

The commands are decoded inside each data path. Commands received from the interface are considered as incorrect when a value or timing error is detected. Value errors affecting the VME interface lead to a decoding error. Timing errors (i.e., no refresh) are detected by checking the maximum time interval between two commands. The time interval between two commands from the host processor is user-configurable and is used as a watchdog signal.

These commands are used to drive the communication lines between the RMBs. The communication lines are implemented by electrical equipotential lines, with one equipotential per pair of channels. A *disconnection request* corresponds to a low level on the equipotential. The implementation is such that a low level is ensured on an equipotential when a channel is not powered. The low level is latched by each RMB to ensure that this equipotential level is removed if and only if both involved channels reset their latches simultaneously.

The RMB voter takes its decisions only on the basis of these equipotential lines. The reciprocity of the *disconnection request* is thus ensured. The decision taken by the voter is a Boolean: an open circuit corresponds to the *disconnection* of the channel (safe state) and a closed circuit to a *connection.*

5.2.2.3 Data Path Comparators

The cross-check of the data path is carried out on the voter decision and the equipotential driver signals. Although the small number of components in each data path ensure a high reliability, a fault affecting one data path may remain dormant until its manifestation on one of the cross-checked signals.

An error is declared if there is a discrepancy in one of the cross-checked signals for a duration longer than the RMB internal asynchronism. The RMB enters a non-recoverable mode, sets its voter output to *disconnect* , drives low its equipotentials and locks this mode until corrective maintenance is carried out.

5.3 Exclusion Logic

The exclusion logic described here provides an alternative to the RMB described in the previous section. In this approach, a faulty channel is irreversibly powered down. Re-insertion of an isolated channel is only possible after a manual re-arming operation.

The *exclusion logic* is a hardware voter, which is distributed over the set of channels and is responsible for excluding faulty channels from the pool. Figure 5.5 illustrates the exclusion logic for a maximal configuration with four channels. The exclusion logic is formed by a distributed arrangement of three types of cards, which are described in the next sub-section. The overall operation of the exclusion logic is presented the second sub-section.

5.3.1 Exclusion Logic Cards

The exclusion logic consists of three types of cards:

- ICN Interface cards (ICN_I)
- Fail-safe AND cards (ANFS)
- Fail-safe DC Power Converter cards (PDCk, $k \in$ {A,B,C})

The required number of cards depends on the level of redundancy (2, 3 or 4 channels). These are specific (non-VME) cards, which are grouped in a proprietary rack that is separate from the VME rack containing the host CPU card(s), the I/O card(s) and the ICN card.

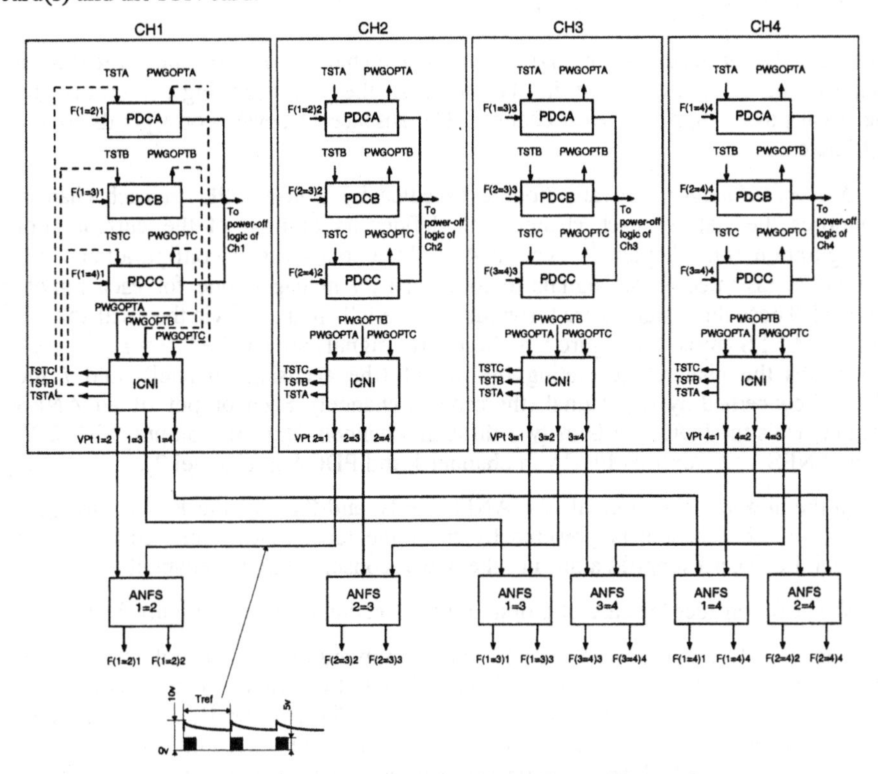

Figure 5.5 — Exclusion Logic for Four Channels

The *ICN Interface cards* (ICN_I) ensure electrical adaptation between the ICN controllers (ICN_C), implemented as piggy-back boards on the ICN manager in the VME rack, the Ethernet transmission lines used to implement the ICN network, and the fail-safe ANFS and PDCk cards. The ICN_I cards are not implemented using fail-safe technology. However, fail-safe operation is ensured by coding permissive states as pulse trains.

The *fail-safe AND cards* (ANFS) are implemented by means of dynamic logic, relying on a transformer that only delivers an output signal when both of its inputs are driven by pulse trains. An incorrect signal (e.g., stuck-at-one), or a failure of one the hardware components of the card, leads to saturation of the transformer and thus to a safe output state. To guarantee that this safe state is irreversible, the card must be manually re-armed by means of a push-button. Such an approach is imposed by standards for safety-critical systems that require: switching to a safe state on first failure; maintenance in the safe state on occurrence of a second (non-simultaneous) failure.

The *DC Power Converter cards* (PDC*k*) consist of DC/DC converters that provide fail-safe amplification of the ANFS outputs. To guarantee safety, the PDC*k* also relies on a transformer, driven by an independent oscillator circuit that is powered directly by the corresponding ANFS output.

5.3.2 Exclusion Logic Operation

The objective of the exclusion logic is to switch a channel to a safe state whenever a majority of channels request it. To do this, the exclusion logic controls the electrical power supply of each channel. The absence of power corresponds to the safe state.

A *Pool Manager* software module executed by a host CPU in each channel diagnoses the state of each channel and defines its opinion of the current pool configuration (cf. Chapter 4, Section 4.2). In a four-channel configuration, each channel sends three on/off signals to its local ICN manager, one for each remote channel. These three signals are forwarded by the ICN manager, via the local INC_I, to three ANFS cards. On Figure 5.5, these three signals are labelled "$i = j$", where i designates the channel generating the signal (the accusing channel), and j the channel concerned by this signal (the accused channel). Each output of an ANFS card controls the PDC*k* cards in two different channels (e.g., the output "F(1=2)1" from ANFS 1=2, controls PDCA in channel 1 and PDCA in channel 2).

In the nominal situation, all the ANFS cards, and thus all the PDC*k* cards, are active, so each channel is powered by the three local PDC*k* cards connected in parallel. Each card supplies a third of the power consumed by the channel.

We now consider the case of a first failure affecting channel 4 (Figure 5.6).

Channels 1, 2 and 3 each send an "off" command (1=4, 2=4 and 3=4), which causes ANFS 1=4, ANFS 2=4 and ANFS 3=4 to switch to the safe state. Consequently, the three PDC*k* cards of channel 4 (PDCA, PDCB, PDCC) remove power from their outputs, switching channel 4 to the safe state. For channel 1 (and similarly for channels 2 and 3), this also causes PDCC to remove power from its output. However, channel 1 remains active, supplied by PDCA and PDCB, each one now supplying half of the power consumed by the channel.

A second failure would lead to a second channel being stopped, with the two remaining channels being powered by a single PDC*k* card. A third failure leads to both remaining channels being stopped simultaneously.

The successive degradation operation described above requires that the PDC*k* cards that remain active be capable of supplying the required power (33%, then 50%, then 100%). An on-line test (inputs TESTA and PoWerGood) checks that each PDC*k* card is capable of supplying 100% of the consumed power. Moreover, the failure of an ANFS card or a PDC*k* card has no effect on pool operation until it is detected by the on-line test of the PDC*k* cards.

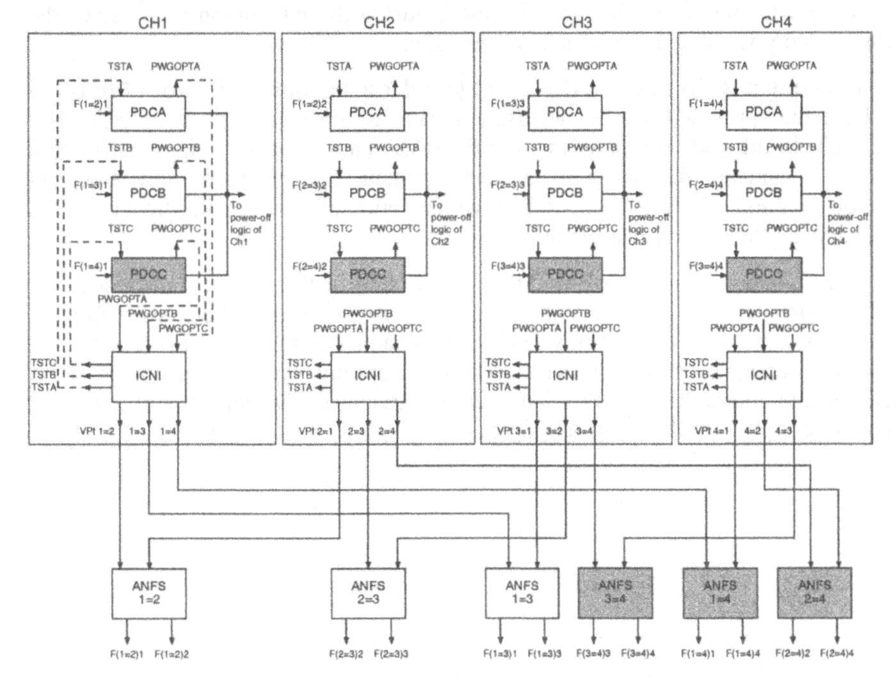

Figure 5.6 — Exclusion of Channel 4

5.4 Conclusion

This chapter has described a generic output consolidation mechanism and two alternative approaches for hardware-implemented channel passivation: channel isolation and channel switch-off using exclusion logic.

The RMB described in the second section of this chapter supports the generic output consolidation and hardware-implemented channel isolation mechanisms. Functional hardware prototypes of the RMB have been developed. These prototypes, which are not fail-safe, are being used for validation and performance measurement.

The RMBs are integrated in the nuclear propulsion prototype (cf. Chapter 10), which is aimed at a safety-related system. The voting algorithm is two-out-of-two with reconfiguration to simplex. Limiting the channel redundancy to two channels was the key point in reducing the overall system cost. The nominal behaviour is a two-out-of-two configuration to ensure the safety objectives. However, to improve the availability of the instance, the nominal configuration can be degraded to a simplex configuration if the faulty channel can be correctly diagnosed. Implementation of this relatively complex voting algorithm was facilitated by the flexibility provided by the integrated chip implementation of the RMB.

Other foreseen operational uses of the RMB may be without output consolidation. The most promising such use is the generation of a hardware reset in the faulty channel. This was partially tested during the integration phase since the RMB prototype already includes this capability.

The current RMB and the ICN manager offer a solution in a two VME-board package that provides all the necessary means to manage a pool of redundant channels. In the mid-term, a redesign of the RMB packaging as a piggy-back board can be considered to take into account of new opportunities with emergent backplane standards like CompactPCI while maintaining compatibility with the widespread VME-based systems market.

The exclusion logic described in the third section of this chapter is intended for fail-safe (irreversible) channel switch-off. In the current demonstrators, switching off the PDC*k* cards does not actually lead to the channel being turned off (instead, an indicator lamp is simply turned off). This option was chosen in order to demonstrate automatic re-integration of an incriminated channel without requiring manual re-arming of the exclusion logic.

Chapter 6

Multilevel Integrity Mechanisms

The purpose of the multilevel integrity mechanisms of the GUARDS architecture is to protect critical components from the propagation of errors due to residual design faults in less-critical components. The notions of multiple integrity levels and multiple criticality levels are very tightly linked, but there is an important distinction. Integrity levels are associated with an integrity policy that defines what is allowed in terms of data flow between levels and resource utilisation by the components at different levels [Totel 1998]. Criticality levels are defined in terms of the potential consequences of failures of components at each level.

The aim of integrity level management is to ensure that low-integrity information does not contaminate high-integrity information. This requires the enforcement of an integrity policy to mediate the communication between software components of different levels. It must also be ensured that it is not possible to bypass the controls put into place to enforce the policy, i.e., illicit communication between application components of different levels of integrity must be prohibited. This is achieved by spatial and temporal isolation.

The most effective spatial and temporal isolation is to have software components of different levels of integrity supported by totally separate machines. In this case, the only means for communication between them is through the network (with adequate temporal firewalls to prevent network hogging). The controls for integrity policy enforcement are then concerned only with network messages. However, this approach is expensive since it requires the use of several machines, even when a single machine is sufficiently powerful to support all the applications.

When several levels of integrity must be supported by a single machine, isolation is more problematic. Spatial isolation requires that each application resides in a different section of memory and is protected by memory management hardware. Temporal isolation must ensure that low-integrity processes do not prevent high integrity processes from accessing essential system resources (such as CPUs or communication paths).

The integrity policy defined in this chapter is based on the notion of integrity levels (similar to the safety integrity levels of IEC [IEC 61508]), and focuses on the

D. Powell (ed.), A Generic Fault-Tolerant Architecture for Real-Time Dependable Systems, 99–119.

problem of the interactions between software components at different integrity levels. Therefore, interactions may occur through sharing of hardware or software resources or by explicit co-operation between entities. Interactions can thus occur either at the system level (the system being composed of the operating system software and the hardware it is running on) or at the application level. Interactions can be divided into two groups: spatial interactions (i.e., modification of code or data) and temporal interactions (e.g., hogging of a resource that becomes unavailable for other application entities). More precisely, each interaction can be either the consequence of a normal operation or the result of a residual design fault in a software component. Our model aims to contain residual design faults. We thus enforce mediation on information exchanges to verify that all information flows during an execution are legal. For this mediation, our integrity policy defines rules on data flows and these rules are enforced by hardware and software mechanisms.

The rest of the chapter is organised as follows. Section 6.1 shows how the integrity policy fits into an object-oriented model, and Section 6.2 gives details of the actual integrity mechanisms. In Section 6.3, the feasibility of the integrity mechanism implementation is demonstrated on a simplified integrity policy, which is currently implemented in an initial prototype. Some other implementations of the policy are sketched out in Section 6.4, while Section 6.5 0presents some related work. Section 6.6 concludes with a summary and discussion of the implementation phase.

6.1 An Integrity Management Model

The proposed integrity management model is based on an object-oriented approach. Our purpose is to define a generic way to characterise and check the information flows between the different objects running in the system. We are then able to prevent the contamination of high integrity objects by lower integrity ones.

6.1.1 Object Model Paradigm

Our model assumes that an object is an entity that provides services that can be requested by clients [OMG 1995]. It is the instance of a class, and is composed of both its internal state (known as its attributes), and its methods (i.e., the services it provides).

The model of invocation is based on message exchange. Each time an object requests a service from another object, it must send it a message which carries both an identification of the invoked method and optionally the data items necessary to complete its execution (the parameters). Each message received by an object results in the execution of the requested service. The execution of a given method can, of course, provoke new invocations, and so the creation of new messages.

Each object is classified within a particular integrity level according to how much it can be trusted (the more trustworthy an object is, the higher its integrity level). The degree to which an object can be trusted depends on the evidence that is available to support its correctness and the consequences that would ensue should it fail (i.e., its criticality). We consider that a message has the same integrity level as the object that created it (we can trust the message contents as much as its creator), and that an integrity label carried by the message itself defines this level. The integrity label enables the integrity rules to be checked at message reception.

6.1.2 Information Flows

Our aim is to avoid error propagation from low integrity level objects to higher level objects. No upward information flow should be allowed since it would be possible for a low integrity object to provide faulty or corrupted information to a high integrity object. This is the main goal of our policy and is the basis of others [Biba 1977]. However, in an object model, all invocations of an object do not necessarily produce a flow of information that can corrupt the invoked entity. Consequently, we will apply constraints on information flows, and not on invocations [Jajodia & Kogan 1990].

Other integrity policies (see Section 6.5) prohibit entirely all information flow from low integrity objects to high integrity objects. However, we find that this approach is too limited. An object can obtain a data item that is of higher integrity than itself, but such a data item will inherit the level of integrity of the receiving object. This results in a degradation of the information's integrity, without any possibility of restoring it. We deal with this drawback by providing objects of a special kind (*Validation Objects*) whose role is to apply fault-tolerance mechanisms on information flows. The purpose of such objects is to produce reliable information from possibly-corrupted data items used as inputs (i.e., with a low integrity level). Such objects upgrade the trustworthiness of data items and allow thereby information flows from low to high integrity levels.

6.2 Integrity Mechanisms

In the following subsections, we describe the integrity mechanisms needed to ensure that no low integrity object can corrupt high integrity objects.

6.2.1 Definitions and Notations

Our intent is to define a flexible and practical integrity policy, which can be easily applicable and usable in a variety of systems and applications. For this reason, we define three kinds of objects:

- **Single Level Objects (SLOs)** are assigned a single constant integrity level, which represents the level of confidence we can have in their execution. Like any object, they can contain information and so create flows carrying their internal data. Basically, they should not read data items from lower integrity levels than themselves, and should not provide information to higher integrity levels.

  However, these rules are very restrictive when building an application: if a service must be made available to many objects with various integrity levels, an instance of the object must be created at every level where it is needed.

- **Multi Level Objects (MLOs)** provide a solution to this problem. They allow more flexibility by being able to provide service to several integrity levels. They are validated at a given integrity level (in the same way as SLOs) and can dynamically modify their current integrity level to be able either to provide a service or use data at the required level. However, they must be designed in such a way that they cannot produce illegal information flows from low to high levels, even between different invocations. To guarantee this property, we choose to ensure that these objects have no memory, i.e., they do not store any state from one invocation to another. This can be done by creating a new object (and thus a new context) at each invocation. Such an object does not keep any memory of previous invocations, and thus acts as a stateless server (e.g., in the same way as NFS [Sun Microsystems 1989]). It can thus be used to implement a service which can be requested at any level.
- **Validation Objects** are particular instances of SLOs that can provide information with a sufficient certified integrity level to allow flows from low to high integrity levels. Such upward flows are essential in many applications. Consider, for example, an airplane guidance control function that needs pressure information coming from unreliable pitot tubes. Sensor data (and the drivers whose role it is to make their values available) are given a low integrity level since they are unreliable. However, the guidance task is given a high integrity level because it is highly critical. As it is currently defined, the information flow policy would prevent the guidance task from reading sensor values. In such systems with unreliable sources, it is common to use several redundant sensors to be able to tolerate the failure of some of them. For instance, the computation of the median of a vector of sensor values will produce a more trustworthy information than any single sensor. In that case, the median computation result can be given a high integrity level, even if each of its inputs is low integrity data. That is the purpose of validation objects: they take low integrity inputs and apply fault tolerance mechanisms to produce high integrity outputs. Consequently, they can be authorised to produce what should be otherwise considered as illegal information flows from a low integrity level to a high integrity level. Of course, these Validation Objects have to be validated at the integrity level of their outputs and must be dedicated to processing specific inputs. Thus, they cannot be polluted by data coming from other low integrity sources.

Formally, we define O the set of objects of the system, which are either single-level objects or multilevel objects, defining thus a partition of O: $\{SLO, MLO\}$. I is the set of the integrity levels and il: $O \rightarrow I$ associates to each object o its *current* integrity level $il(o)$. We also define a partial order relation ($\leq$) between the elements of I.

6.2.2 Integrity Policy

We now describe our integrity policy and how it is applied to control information flows. The integrity architecture we propose is described in Figure 6.1.

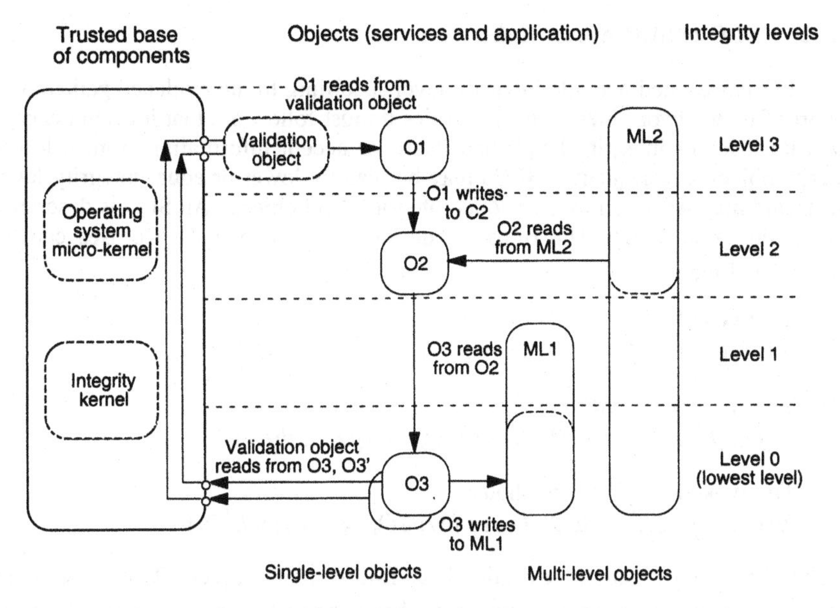

Figure 6.1 — Integrity Architecture

6.2.2.1 Method Classification

Two sorts of information flow can occur: *explicit* data flow (which implies application data flow, i.e., attributes passing through an object method invocation) and *implicit* data flow, (i.e., when the execution of the called method can potentially have a consequence on the object state). In both cases, a flow of information sent to a method can corrupt it. To control data flows, we must be able to determine whether a call to a method is carrying information or not. We thus distinguish three types of methods:

- **Read invocation**: the invocation provokes no state modification of the called object. Typically, this corresponds to a request to obtain information from an object. There is an information flow from the invoked object to the caller.
- **Write invocation**: the invocation provokes a modification of the state of the called object but no modification of the caller internal data. There is thus an information flow from the caller to the called object. This information flow can either be *explicit* (the method has entry parameters) or *implicit* (for example, requesting an object to increment one of its internal attributes).
- **Read-Write invocation**: this type of method generates a bi-directional flow between the calling and the called objects.

This classification is the basis of the rules we define in the next section. According to the type of a given method, we can evaluate if an invocation could corrupt an object, and decide if an invocation is permitted or not, depending on the integrity levels of the calling and called objects.

6.2.2.2 SLO to SLO Access Rules

Each SLO is assigned a single integrity level according to the evidence gathered in support of its trustworthiness. The only rules it must follow are that it cannot access data with a lower integrity level than itself, or accept information from a lower integrity object. Consequently, SLOs may be read by lower or equal integrity level objects and may write to lower or equal integrity level objects. An SLO is described using a singleton $il(O)$. The rules we define on SLO to SLO accesses can be described as follows:

- To invoke Read methods:
  $\forall(O_1, O_2) \in SLO \times SLO,\ O_1\ read\ O_2 \Rightarrow il(O_1) \leq il(O_2)$
- To invoke Write methods:
  $\forall(O_1, O_2) \in SLO \times SLO,\ O_1\ write\ O_2 \Rightarrow il(O_1) \geq il(O_2)$
- To invoke Read-Write methods:
  $\forall(O_1, O_2) \in SLO \times SLO,\ O_1\ readwrite\ O_2 \Rightarrow il(O_1) = il(O_2)$

The three rules can be explained as follows: an object O_1 can read an information from an object O_2 only if O_1 has at most the same integrity level as O_2, i.e., if O_2 can provide information with a sufficient level of integrity, so that O_1 cannot be corrupted. Additionally, an object O_1 can provide information to an object O_2 (write rule) only if it has at least the same integrity level as O_2, i.e. it can provide data which are reliable enough to be used by this object. Finally, read-write operations (i.e., information exchange) can be completed only if both objects can have mutual confidence. The read-write rule is in fact defined as a logical **and** between the read and write rules.

6.2.2.3 MLO Access Rules

Like a single level object, an MLO is also assigned an integrity level, according to the evidence gathered in support of its trustworthiness. This level is called its *intrinsic* level. But unlike a single level object, an MLO is able to run at any level less than or equal to its intrinsic level. Since it is necessary to isolate different invocations which execute concurrently within an MLO, we require that the MLO have no memory (i.e., no persistent state used from one invocation to another). Due to our basic requirements for no upward information flow, an MLO can downgrade its integrity level during a method execution to reflect the level of information it uses, but is not able to upgrade it until the full completion of this execution of the method.

Formally, we describe an MLO using a triple, indexed by an invocation identifier α, $\{maxil(O), il_\alpha(O), minil_\alpha(O)\}$, where: $maxil(O)$ is the intrinsic level which can be reached by the MLO, $il_\alpha(O)$ the current integrity level (which reflects the level of information used so far during the execution corresponding to invocation α), and $minil_\alpha(O)$ the minimum integrity level which can be reached by the object during its execution. The latter allows an optimisation with respect to read invocations that could not be completed due to an integrity conflict (i.e., an early detection of an illegal flow resulting from the read invocation). At any time, $il_\alpha(O) \in \left[minil_\alpha(O), maxil(O)\right]$.

The rules that apply to an MLO have to take into account the basic rules (i.e., no upward information flow) and consider the ability of an MLO to downgrade its level to adapt itself to a given integrity level. For example, if an MLO O reads a data item of level il_{data}, its current integrity level $il(O)$ will reflect this level (only if $minil_\alpha(O) \le il_{data}$, i.e., the MLO has permission to access the data item). Then it will not be authorised to write this information to a higher level (i.e., the SLO write rule must be applied on its current level).

The MLO to SLO access rules can be defined as follows:

- Read invocation:
  $\forall(O_1, O_2) \in MLO \times SLO, \quad O_1 \; read \; O_2 \Rightarrow minil_\alpha(O_1) \le il(O_2)$
  After execution, the read invocation condition is still valid, but the MLO level must be updated, i.e., $il_\alpha(O_1) \leftarrow \min(il_\alpha(O_1), il(O_2))$

- Write invocation:
  $\forall(O_1, O_2) \in MLO \times SLO, \quad O_1 \; write \; O_2 \Rightarrow il_\alpha(O_1) \ge il(O_2)$

- Read-Write invocation:
  $\forall(O_1, O_2) \in MLO \times SLO, \quad O_1 \; readwrite \; O_2 \Rightarrow minil_\alpha(O_1) \le il(O_2) \le il_\alpha(O_1)$
  After execution, the MLO level must reflect the level of the information received, i.e., $il_\alpha(O_1) \leftarrow \min(il_\alpha(O_1), il(O_2))$

We can note that the return of a read invocation by an MLO implies the update of its current integrity level, which demonstrates its ability to adapt its level.

In the same way, the SLO to MLO access rules can be described as follows:

- Read invocation:
  $\forall(O_1, O_2) \in SLO \times MLO, \; O_1 \; read \; O_2 \Rightarrow il(O_1) \le maxil(O_2)$.
  If this condition is verified, we must then ensure that O_2 cannot provide data with a level lower than $il(O_1)$ by updating $minil_\alpha(O_2)$, i.e., $minil_\alpha(O_2) \leftarrow il(O_1)$. Moreover, as a read invocation does not produce an information flow from O_1 to O_2, O_2 can be used at its highest level, i.e., $il_\alpha(O_2) \leftarrow maxil(O_2)$.

- Write invocation:
  Such an invocation is always permitted (the MLO can adapt its level to any level less than or equal to its intrinsic level), and acts upon the current integrity level of the invoked object, i.e., if $(O_1, O_2) \in SLO \times MLO$, and O_1 writes to O_2, then $il(O_2) \leftarrow \min(il(O_1), maxil(O_2))$.

- Read-Write invocation:
  $\forall(O_1, O_2) \in SLO \times MLO, \; O_1 \; readwrite \; O_2 \Rightarrow il(O_1) \le maxil(O_2)$.

The MLO must not be able to provide information with an integrity level lower than $il(O_1)$, and its current level must reflect the information provided by the SLO, i.e., $minil_\alpha(O_2) \leftarrow il_\alpha(O_2) \leftarrow il(O_1)$.

The MLO to MLO access rules are very similar to the SLO to MLO access rules, but take into account the particular properties of the multilevel objects. In particular, we can remark that during the execution of an invocation, the current integrity level of an MLO can only downgrade.

- Read invocation:
  $\forall(O_1, O_2) \in MLO \times MLO,\ O_1\ read\ O_2 \Rightarrow minil_\alpha(O_1) \le maxil(O_2)$.
  If this condition is verified, we must then ensure that O_2 cannot provide data with a level lower than $il(O_1)$ by updating $minil_\beta(O_2)$, i.e., $minil_\beta(O_2) \leftarrow il(O_1)$. Moreover, as a read invocation does not produce an information flow from O_1 to O_2, O_2 can be used at its highest level, i.e., $il_\beta(O_2) \leftarrow maxil(O_2)$.
  After execution, the read invocation condition is still valid, but the calling MLO level must be updated, i.e., $il_\alpha(O_1) \leftarrow \min(il_\alpha(O_1), il_\beta(O_2))$.

- Write invocation:
  Such an invocation is always permitted (the MLO can adapt its level to any level less than or equal to its intrinsic level), and acts upon the current integrity level of the invoked object, i.e., if $(O_1, O_2) \in MLO \times MLO$, and O_1 writes to O_2, then $il_\beta(O_2) \leftarrow \min(il_\alpha(O_1), maxil(O_2))$.

- Read-Write invocation:
  $\forall(O_1, O_2) \in MLO \times MLO,\ O_1\ readwrite\ O_2 \Rightarrow minil_\alpha(O_1) \le maxil(O_2)$.
  The MLO must not be able to provide information with an integrity level lower than $il(O_1)$. Its minimum level must reflect the minimum level of the invocation, i.e., $minil_\beta(O_2) \leftarrow minil_\alpha(O_1)$. Its current level must also reflect the information provided by the SLO, i.e., $il_\beta(O_2) \leftarrow \min(il_\alpha(O_1), maxil_\beta(O_2))$. After execution, the MLO level must reflect the level of the information received, i.e., $il_\alpha(O_1) \leftarrow \min(il_\alpha(O_1), il(O_2))$.

These rules are described in detail in [Totel 1998].

6.2.2.4 Validation Objects

Validation Objects must be designed in the application at the needed level of integrity (cf. Figure 6.1) with the development and validation methods appropriate to that level. Generally speaking, the validation objects should be as simple as possible, and should implement only the fault tolerance mechanisms required to successfully validate the data. Keeping such objects as simple as possible should facilitate their validation. The information flow from specific low level inputs to a validation object must be considered as an exception to the integrity policy rules. The exceptions are identified and defined by the application programmer, and then authorised by the integrity kernel (i.e., the software entity responsible for the integrity checks).

6.2.3 Firewalls

To ensure that GUARDS application processes of different criticality do not interfere with one another, processes are restricted not only by different integrity levels but also through spatial and temporal protection mechanisms. Spatial firewalls are necessary to protect the application's address space while temporal firewalls are needed to allocate resource occupation times to each process or thread.

6.2.3.1 Spatial Firewalls

Spatial protection of a process is usually achieved with memory management hardware and software. A minimum set of features must be provided at the hardware level to support the implementation of a policy at the software level. Some policies implemented in memory management units (MMUs) can be quite complex: for example, the Intel 80x86/Pentium processors can handle four levels of memory protection and apply a ring level protection policy on the memory segment accesses (the rules it applies are very different to those of our integrity policy).

The memory protection must provide a strict isolation between process memory segments to ensure that information flow control between processes can not be bypassed. This basic functionality is part of every protection mechanism provided by memory management units of the various processors chosen for the GUARDS prototypes.

6.2.3.2 Temporal Firewalls

Even with spatial firewalls and the GUARDS integrity policy for object communication, it is still possible for a low criticality process to affect high criticality processes by consuming more resources (e.g., CPU execution time, bus occupation time...) than anticipated. This is particularly a problem where the low and high criticality processes share the same infrastructure. To facilitate a predictable temporal behaviour of a GUARDS application, both computational and scheduling models have to be defined. An application conforming to the GUARDS computational model consists of periodic and sporadic processes and the mechanisms for their communication. Three computational models may be defined (see Chapter 3): a) time-triggered, b) event-triggered, c) mixed (time and event-triggered). The time-triggered computational model consists of a fixed set of periodic processes that are released by a timer event. In the event-triggered model, sporadic processes are released by an event originating from either another process or an interrupt. The computational model chosen for GUARDS is a mixed computational model consisting of periodic and sporadic processes, which communicate asynchronously through shared memory.

A scheduling model dictates the way in which the processes of an application conforming to a particular computational model are executed at run-time. The runnable processes are executed in the order determined by their priority. In the GUARDS pre-emptive priority scheduling model, if a higher priority process becomes runnable (e.g., it is released by a timing event), then the current lower priority process is stopped from executing and the high priority process is executed. The lower priority process executes again when it is the highest priority runnable process. Pre-emptive priority scheduling automatically provides a limited form of temporal firewalling between processes with different priorities. However, the assignment of a priority to a process is not solely a result of the process's criticality. Rather, deadline monotonic priority ordering (for example) assigns priorities according to the deadlines. Hence, it is possible for a low criticality process to have a short deadline and therefore potentially a high priority. To solve this problem, either the priority range needs to be divided and/or CPU budget timers need to be supported [Burns & Wellings 1995a]. In the time-triggered computational model, temporal firewalls between different processes require the use of budget timers. A CPU budget timer is set to the amount of CPU time required by a process for each invocation. Whenever the process is pre-empted, the timer is stopped, and is restarted next time it executes. In the event-triggered computational model, the operating system kernel must maintain a sporadic process' invocation rate and budget timer to ensure the temporal firewalls between the processes.

6.3 Implementation

The integrity policy defines how the interactions between the different software components in a system must be checked to ensure that no highly critical component can be corrupted by a possibly faulty low integrity component. As stated above, these interactions can appear at many different system levels. The integrity policy we defined in Section 6.2 can thus have an impact at many different system layers.

6.3.1 Impact of the Policy at Different System Layers

There are many hardware or software loopholes at various levels in an execution environment that can allow a faulty object to corrupt another object. To be able to apply our policy to all possible information channels at different layers, we need to interpret our policy in terms of current system architectures.

Basically, the object model encapsulates both the data items and the methods necessary to execute the service. An object can modify the internal state of another object by calling an execution entry point corresponding to a method. In our model, the contexts of the different objects are private to each object and can be modified only by the methods encapsulated within the object itself.

6.3.1.1 Hardware Layer

We could consider implementing our policy at the hardware level. In this case, only the process that is allocated a data segment should access it. Each entry point in the code segment (called a Call Gate in the Pentium MMU) should be defined and integrity rules applied to any call through these entry points. Some MMU improvements could be envisaged so that a code segment could read data segments with an integrity level greater than or equal to its level, and write into segments of equal or lower integrity. Although valuable and of interest, such a hardware implementation is not realistically feasible in a COTS-oriented approach. Instead, the GUARDS approach is to verify that the COTS MMU can provide a full spatial isolation of the different processes (see Section 6.3.2).

6.3.1.2 System Layer

In addition to memory management, an operating system provides the application layer with different communication channels. Once sufficient isolation between the processes (defined by their code and data segments) is ensured, the communication primitives provided by the operating system kernel are the only way for the processes to exchange information. These communication functions can have proprietary or standardised interfaces (particularly POSIX communication interfaces) that we must adapt to be able to apply integrity rules on the information flows. An integrity kernel integrated in the operating kernel is then in charge of verifying which information flows are permitted or not. This approach requires access to the kernel source code (which is not the case in most situations) but has the advantage of ensuring that the integrity checks cannot be bypassed by the applications. We have completed an implementation of this approach in the Chorus[1] micro-kernel [Totel 1998], but it can be adapted to any operating system (such as QNX[2,3] and VxWorks used in the GUARDS project).

6.3.1.3 Middleware Layer

In a COTS-oriented approach, modifying the operating system kernel is not appropriate. The only way to have reusable components on many platforms is to implement the integrity checks in a layer of software that lies above the operating system kernel. The idea is that all communication between application processes goes through this middleware layer so that they can take advantage of the integrity mechanisms. There are two ways in which this can be achieved:

1. By a middleware component that must be called *explicitly* by the application every time it wishes to communicate.
2. By a middleware component that is called *implicitly* (in a transparent manner) by linking to an alternative library to that which would normally be used to call the kernel. This technique was used in the Newcastle Connection [Brownbridge *et al.* 1982]: all system calls can be intercepted, code executed to check that the integrity rules have been adhered to, and then the appropriate trap to the kernel performed.

The latter approach has been adopted in GUARDS since it enables integrity level checks to be transparent to the application.

---

1 Chorus is a trademark of Chorus Systèmes, now Sun Microsystems [Rozier *et al.* 1990]

2 QNX is a trademark of QNX Software Systems Ltd.

3 VxWorks is a trademark of Wind River Systems.

6.3.2 Support for Firewalls

We describe here how the operating systems chosen for GUARDS support spatial and temporal firewalls, and integrity level management.

6.3.2.1 Choice of Operating Systems

The key requirements for a GUARDS operating system are: (a) it should be a COTS product, (b) it should be POSIX compliant [IEEE 1003.1] and (c) it should provide the facilities required by the GUARDS computational and scheduling model (e.g., clock access, support for different process priority levels, scheduling algorithms, interrupt handling, pre-emption), and for spatial and temporal firewalling (e.g., memory management, CPU timers). More specifically, a typical GUARDS application requires:

1. Basic operating system services (process creation and deletion, signals, basic I/O, and terminal handling),
2. Real-time services (clocks and timers, process scheduling control, real-time signals, semaphores, shared memory, message queues, asynchronous I/O),
3. Thread services providing the ability to run multiple, concurrent threads within a single address space, since processes in POSIX execute in separate address spaces (pthreads, mutexes and condition variables).

The facilities provided by POSIX support the GUARDS computational and scheduling model as they provide the functionality needed for periodic and sporadic processes, communication and synchronisation between processes, and pre-emptive priority based scheduling. The two operating systems chosen for GUARDS applications are VxWorks and QNX. As both operating systems are compliant only to POSIX 1003.1 and 1003.1b, at the start of the implementation stage it was decided to use the POSIX service if it is implemented, otherwise to use the equivalent proprietary service.

6.3.2.2 Spatial Firewalls

Spatial firewalls exist to prevent a low criticality process from accessing the virtual memory space of a high criticality process. POSIX provides suitable mechanisms for spatial firewalls in the form of process abstractions. Different processes need to be mapped to separate address spaces so that the memory management facilities provided by the host processor ensure spatial firewalling. However, VxWorks does not support processes. Furthermore, MMU hardware and software are optional in VxWorks. Hence, the trust of the spatial firewalls implementation in VxWorks lies on its architecture-independent interface to the CPU's memory management unit in the form of the virtual memory support option VxVMI. In QNX the operating system kernel fully utilises the MMU to deliver the POSIX multi-process model.

6.3.2.3 Temporal Firewalls

Temporal firewalls should prevent a process from using the processor or a system resource for longer than has been allocated. To ensure that a process does not exceed the processor usage that it reserved, the processor usage of each process must be tracked. Watchdog timers can be created from the POSIX timer mechanisms, so that deadline overrun can be detected. However, CPU budget timers are currently not defined by the POSIX standards (although they are under consideration). Consequently, it is not possible to confine an error resulting from one process consuming more CPU time than has been allocated to it. The inevitable result of this poor error confinement is that damage assessment is difficult when a deadline overrun occurs.

6.3.3 Description of the Spatial Firewall Implementation

For portability, the implementation has been carried out in a middleware layer, by designing a specific communication library. The goal of this component is to implement a simplified version of the integrity policy that has been defined. In particular, to reduce the complexity of the checking mechanisms, only Single Level Objects are used. Each GUARDS application is considered to be an SLO running in its associated integrity level. Validation Objects are implemented by defining exceptions on the information flows between the interacting SLOs (so that specific certified upward flows can be permitted).

To implement the checking mechanisms, two sub-components (see Figure 6.2) are defined:

- A library that must be linked to each application, and whose role is to supply the specific GUARDS communication routines to the application.
- An integrity kernel, whose role is to perform all checks on the information flows produced between the applications.

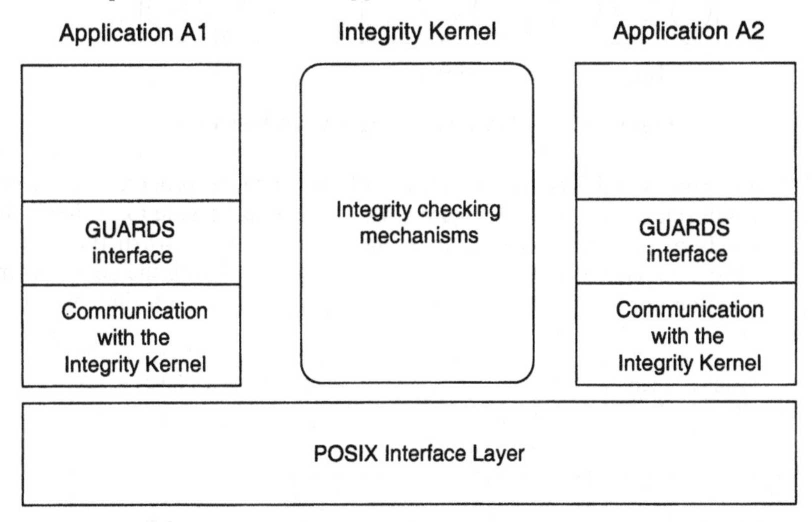

Figure 6.2 — Integrity Mechanism Components

The GUARDS interface is based on a POSIX layer, so to as to be portable between the various GUARDS implementations. Basically, the role of this interface is to control the use of the following POSIX functionalities: message queues, signals, shared memory, semaphores, pthread mutex and memory locking.

In the operating systems selected for GUARDS, the notion of an object does not exist. The active entities are processes, and are not invoked directly (except if only RPC-like mechanisms are used, which is quite restrictive). Consequently, the object-oriented invocation scheme defined in the integrity policy has been interpreted to adapt it to a conventional client-server system, and to handle active and inactive entities.

6.3.3.1 Integrity Kernel Description

The distributed integrity notion we use stems from the security kernel approach defined by the National Computer Security Center (NCSC) in the red book [NCSC 1987] (see Figure 6.3). Each integrity kernel is responsible for checking accesses to local objects (local-to-local as well as distant-to-local accesses). An important point to note is that in this approach, each integrity kernel trusts the other kernels hosted on other sites. If an object O_1 requests access to an object O_2, the kernel at the first site sends the request to the second site. It also builds an integrity label containing the information items necessary to check the access. This label will be carried by the message sent to the distant site.

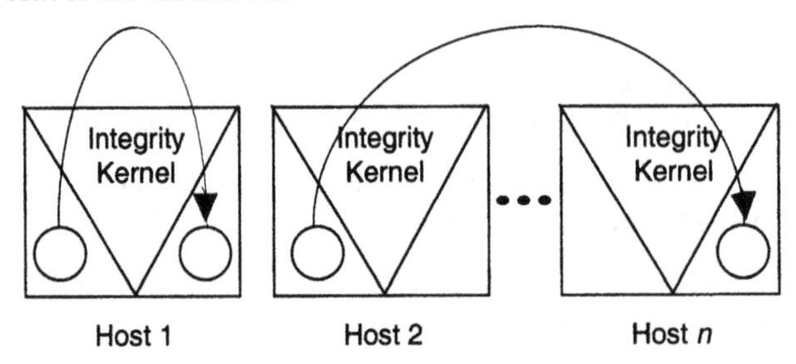

Figure 6.3 — Red Book Integrity Architecture

Each integrity kernel has the knowledge of the integrity properties of every object that will run on its site (see Figure 6.4). It will thus be able to initialise the integrity levels of the different objects at their creation (i.e., when the objects register with the integrity kernel). The definitions produced during the development of the application are stored statically and used by the kernel each time an object registers.

Object Id	Integrity Level	Method ID	Method mode	Object key	Channel Id	Invocation_ queue Id
Id_O	1	ID_m	read	Key_O	CH_O	Id_{iy}

Figure 6.4 — Integrity Kernel Data Structure

The information available statically describes:

- The identifier of the object.
- The level of the object.
- The public methods that can be invoked by other objects, and their modes.
- The access exceptions induced by the use of Validation Objects.
- Message queues descriptors for the integrity kernel to communicate with the object.

6.3.3.2 Library Description

A specific library is supplied to all applications. This library provides three groups of specific routines to allow applications to communicate with other applications:

1. Routines that must be used by all objects in the system:
   - *guards_object_register:* this routine must be used by every object in the system so that it can be registered as a GUARDS application object (this information is stored by the integrity kernel). If this routine is not executed, the object will not be authorised to communicate with other GUARDS application objects.
   - *guards_object_unregister*: this routine allows a GUARDS application object to unregister before exiting.
2. Routines that must be used by server objects (which provide routines that can be invoked by clients):
   - *guards_receive*: this routine allows an application object to receive invocations from other application objects. This routine returns information about the method which is invoked and a descriptor of the invoking object.
   - *guards_reply*: this routine allows a server to reply to an invocation of one of its methods. A call to *guards_reply* necessarily follows a call to *guards_receive* and uses the object descriptor returned by *guards_receive*
3. Routines that must be called by client objects (objects that invoke routines of other objects):
   - *guards_call*: this routine allows an application to invoke a particular method of an other application. To make this invocation, the objet must know the object_id of the target object and the method_id of the invoked method. A *guards_call* may be either a blocking call or a non-blocking call
   - *guards_get_reply*: this routine is used in case of a non-blocking *guards_call*. As a matter of fact, if an object invokes a method of another object as a non-blocking call, the object may execute some other piece of code before checking if the reply from the invoked object is received through the *guards_get_reply* routine.

6.3.3.3 Communication Library

The communication library is responsible for managing all the communications between each application object (through the GUARDS specific library) and the integrity kernel. These communications are of course transparent from the application's point of view. When an application uses one of the routines of the GUARDS specific library, the communication library sends information to the local integrity kernel indicating what kind of operation the object wants to perform. The communication library also receives all messages from the integrity kernel to the local objects. Communication is implemented using POSIX message queues.

6.3.3.4 Structure of the Integrity Kernel

The integrity kernel is a sporadic object that starts two active threads. One thread constantly reads messages in a particular message queue and then executes operations according to the type of messages received. The messages received on this message queue are all local messages sent by local objects. The second thread constantly reads messages that come from the ICN (see Chapter 2) and executes operations according to the messages received. The messages received are all remote messages that come from other integrity kernels.

The integrity kernel first starts by initialising the data structure with the characteristics of the local applications (some information is also stored about remote applications but concerning essentially localisation rather than integrity). Then it creates the two threads described above.

6.3.3.5 Integrity Checking Mechanisms

Integrity checking is carried out by collaboration between integrity kernels (see Figure 6.5). When an application object O_1 wants to invoke a method of a remote application object O_2, the integrity kernel of O_1 sends a message to the integrity kernel of O_2 and adds integrity information about O_1 in the message. The integrity kernel of O_2, knowing the integrity level of O_2 and using the integrity information about O_1 inserted in the message, can make integrity controls and then authorise or refuse the call.

6.4 Other Implementations

For research purposes, the integrity mechanisms have also been implemented on top of two other systems: an object-oriented CORBA compliant system and a Chorus micro-kernel. This section will briefly describe the implementations and the results obtained.

6.4.1 CORBA Implementation

Several parameters dictated the choice of CORBA [OMG 1995] to demonstrate the feasibility of the mechanisms defined by our policy. First, our model is based on an object model that is very close to that of CORBA. The natural notion of distribution in such an environment facilitates the implementation of a distributed integrity kernel. Moreover, each object in such an environment owns a context that is isolated from the other objects (spatial isolation). Temporal isolation cannot be implemented on top of a CORBA layer, but we assumed this would be achieved at operating system layer. Finally, thanks to recent evolutions in object technology, the reflexivity notion [Kiczales *et al.* 1991] was fully applicable and usable to ensure that the integrity checks are transparent and cannot be bypassed.

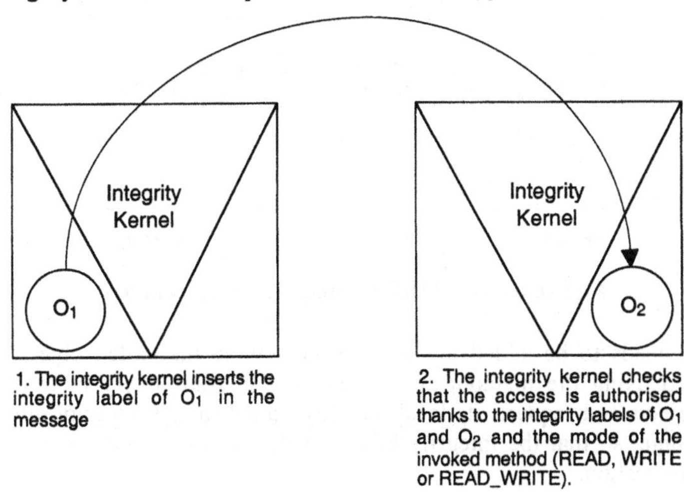

Integrity
Kernel
O_1
Integrity
Kernel
O_2
1. The integrity kernel inserts the integrity label of O_1 in the message
2. The integrity kernel checks that the access is authorised thanks to the integrity labels of O_1 and O_2 and the mode of the invoked method (READ, WRITE or READ_WRITE).

Figure 6.5 — Integrity Checking Mechanisms

This implementation can also have a real significance in wide area networks, where heterogeneous systems are interconnected via an Object Request Broker (ORB). It could be used to prevent the contamination of local critical applications by external unknown software entities. Particularly, we could easily imagine applications on Internet where Java applets can use the CORBA communication interface.

6.4.1.1 Global Architecture

The distributed architecture we assume is described in Figure 6.6. Application objects are executed on several distributed hosts and can communicate via the ORB. Each machine hosts a local integrity kernel whose role is to check the method invocations.

In a standard middleware approach, all functions necessary for communication must be enclosed in a library. These functions will then appropriately call the integrity kernel object. One of the main problems with this approach is that the programmer has to explicitly call the integrity-related communication primitives, otherwise his/her application will bypass the integrity checks.

In object-oriented development, the use of metaobjects has been shown to be a valuable concept for implementing fault-tolerance, providing transparency and reusability [Fabre & Pérennou 1998]. We followed this approach by encapsulating the integrity functions in a metaobject whose role is to trap all method invocations. This metaobject then calls the integrity kernel, which permits or refuses the execution (Figure 6.7). Depending on the result, the metaobject will continue or abort the method execution. At the end of method execution, the same process is applied on return from the invocation.

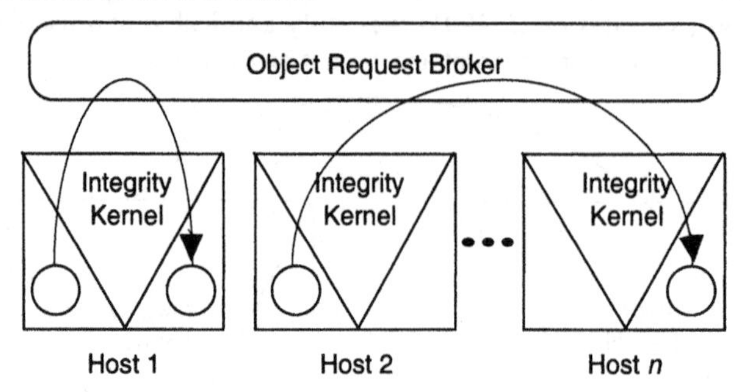

Figure 6.6 — CORBA Integrity Architecture

A label needs to be added to the message sent to the called object. This is implemented using the metaobject. The latter also traps every method call to CORBA objects and requests its local integrity kernel to add integrity information to the message, so that the integrity kernel of the destination host can apply the checks on invocation.

We used Open-C++ V2 [Chiba 1995] for this implementation. It generates a pre-compiler that encapsulates all metaobject definitions. The programmer has just to use this pre-compiler and declare that every object must be supervised by the integrity metaobject.

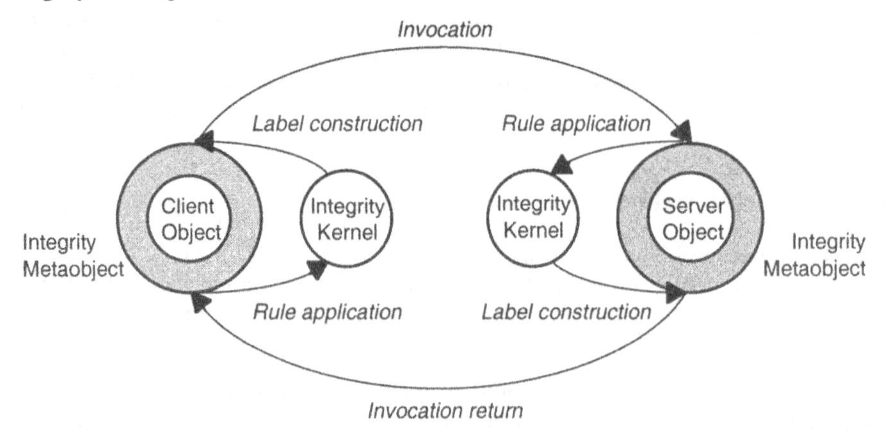

Figure 6.7 — Integrity Kernel Invocation Checks

6.4.1.2 Results

One of the main reasons for this implementation was to test the efficiency of the integrity checks. We measured several time parameters on a sample of 50 000 invocations. The average measures obtained on a Sun Sparc station 20 are summarised in Table 6.1. The time overhead is composed of additional communication time due to the messages that invoke the integrity kernel services and the time necessary to check the client inside the kernel. The comparison between a checked invocation (Integrity Kernel active) and a normal one (Integrity Kernel inactive) was based on the delay from call to return as seen from the client. The result was that an invocation checked by the integrity kernel is about 4 times slower than a normal one. However, the time needed to perform all checks inside the kernel (invocation plus return of invocation checks) is not significant compared to a normal invocation time (0.02 ms compared to 0.62 ms). The bulk of the overhead is thus due to the metaobject trapping mechanisms and the communication delays they induce. So, if the checks were directly inserted into the Object Request Broker (this would be a more efficient implementation in a CORBA middleware layer), the integrity policy check times could be considered as negligible.

Table 6.1 — Invocation Time

Integrity Kernel inactive		**Integrity Kernel inactive**
Total invocation time	Total checking time	Total invocation time
2.44 ms	0.02 ms	0.62 ms

6.4.2 Chorus Implementation

The purpose of the implementation in a micro-kernel was to demonstrate that the approach could be fully and efficiently mechanised. It aims to demonstrate that all identified information flows can be controlled inside the system itself, extending thus the checks to all software entities including system services. As the checks are performed by the kernel itself, no application can bypass them.

However, such an implementation requires an interpretation of the object model on which our integrity policy is based:

- The active object definition we used until now is not supported by the kernel. Consequently, the software components must be divided in two classes: active entities (actors in Chorus), and passive entities (such as files). Moreover, the integrity checks must be applied both on the communication primitives used by the actors, and on the kernel functions that provide access to the passive entities.

- The notion of a method does not exist, and thus must be interpreted: we consider that a method is an entry point in a program. This entry is provided for use by inter-process communication primitives.

The integrity kernel is similar to that described previously. Its role is to provide primitives for initialising integrity levels and for carrying out integrity checks. The modification of the micro-kernel consists in inserting in the integrity kernel source code calls to these primitives, so that all information flows can be checked. To reach this goal, we must ensure that:

- No direct information flow is possible (for example, by direct use of shared memory mechanisms).

- Each time an actor or a passive object is created, correct integrity information is assigned to these entities (see Figure 6.8).
- All communication primitives are trapped to check the integrity labels associated with the interacting object. As in the previously described implementations, the integrity kernel is distributed and performs the checks by using the integrity labels assigned to each message.

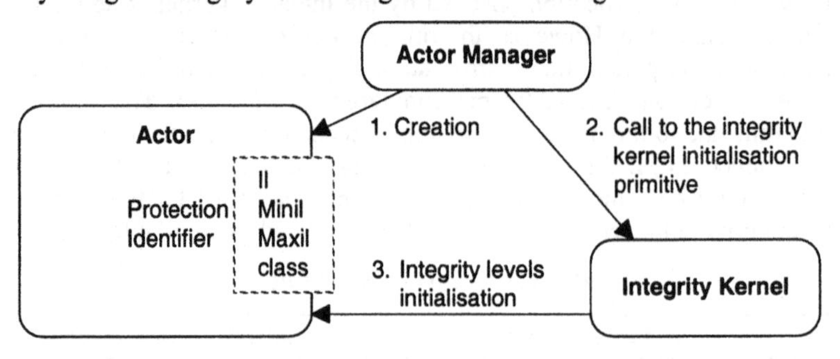

Figure 6.8 — Actor Creation

6.5 Related Work

Biba defined a multilevel integrity policy based on the definition of a one-way information flow in a lattice of security labels [Biba 1977]. The goal of this policy is to prevent access by subjects (active entities including users, processes) to data items (passive entities called objects) and component services that would be illegal from an integrity point of view. The checks defined by Biba are concerned with all kinds of data flows (observation, modification and invocation of components). The rules used for these checks can be summarised as follows: an object can be read by a subject only if the integrity level of the subject is less than or equal to the object integrity level; an object can be modified by a subject only if the object integrity level is less than or equal to the subject level; a subject can invoke another subject only if the latter has an integrity level that is less than or equal to the former. This induces a degradation of information integrity (see Section 6.1.2) which can be corrected by means of a "trusted process" which has to be part of the Trusted Computing Base.

The Clark and Wilson integrity policy [Clark & Wilson 1987] was built for commercial purposes and focuses on the preservation of data integrity. Each data item can be modified only through the use of a certified procedure whose role is to process the data item. This guarantees that the data item will keep its integrity. Moreover, the data items can be verified by dedicated procedures whose role is to certify their integrity. Unlike Biba's policy, the integrity rules focus on the correctness of data modification instead of information flow control.

Kopetz presents an interesting but very different notion of temporal firewall that aims to avoid temporal locks between tasks that co-operate [Kopetz 1997]. It is defined as a data-sharing interface: the data items it provides are the results of the subsystem execution behind the firewall and are certified to be temporally accurate at their time of use. A producer can make information available to the consumer by way of the firewall. The consumer can then access the data items when needed in the time interval defined for this purpose. This temporal firewall can be considered as an error containment interface since data errors can only propagate from the producer to the consumer through a one-way communication channel. It could be used in GUARDS to prevent temporal interactions between objects.

Boeing has built an architecture, called Integrated Modular Avionics (IMA), that provides spatial and temporal isolation as well as error confinement properties. The isolation is based on the notion of a partition. Such a partition supports the execution of a collection of software processes. The partitions defined in the system have separate memory spaces that may be shared with other partitions, but not written by them, in order to prevent data or code contamination. Moreover, an error that occurs in a partition cannot affect other partitions (complete mechanisms are implemented to separate executions in a partition from other partition executions, such as the saving of all processor registers used by a given partition). The spatial isolation is implemented using a memory management unit, and time-partitioned access to shared resources through SAFEbus [Hoyme & Driscoll 1992]. The temporal isolation is based on deterministic scheduling (each partition has the guaranteed access to a needed hardware resource in a given time slot) and is implemented through SAFEbus partition synchronisation. The execution isolation of processes is more complete than in GUARDS but is not based on a COTS approach. However the access rules defined in our approach are more complete than in IMA access policy.

6.6 Conclusion

Integrity management can influence the different architecture layers. In some approaches (such as in IMA) the checks rely mainly on the lower layer, i.e., hardware. In GUARDS, the use of COTS hardware involves implementing the checks in the operating system or in a middleware layer. But this requires that strong hardware properties are verified (e.g., a full spatial isolation). A more complete approach would be to use the complementarity of both hardware and software layers to provide a homogeneous control of the interactions at every level of the architecture. In particular, temporal aspects of the integrity management can be integrated in middleware at the cost of some loss in performance (e.g., the use of timers to prevent tasks from hogging CPU is not very effective). However, the current implementation of integrity checks inside the Chorus operating system proved that the complementarity of an operating system and middleware approach is sufficient to fulfil our requirements (as long as a sufficient isolation is performed in hardware). On the other hand, a complete implementation in a middleware layer, on top of a COTS operating system, would not prevent the existence of covert channels inside the operating system itself.

Chapter 7

Architecture Development Environment

A coherent and rigorous methodology must be applied during the design and development of dependable real-time software systems, with the support of appropriate tools. This chapter describes the Architecture Development Environment (ADE) designed to support the GUARDS methodology through a set of integrated tools for the consistent management of all dependability information and properties, and at the same time offering specific features which address issues of fault-tolerant architecture.

The ADE consists of a set of tools for designing instances of the GUARDS architecture. The tool-set allows collection of the performance attributes of the underlying execution environment and the analysis of the schedulability of hard real-time threads, not only within each processing element of the system, but also among them. This allows in particular a rigorous definition of critical communication and synchronisation among the redundant computers.

The supporting tools are applied throughout the development process, to enable an early verification of performance requirements during the design phase and their continuous assessment during other development phases. A consistent mapping of the design into code is necessary, in order to preserve the achieved verification results in the system actually implemented. These tools allow the application designer to:

- Elaborate an instance of the GUARDS architecture from a set of building blocks.
- Design the application software on top of the defined instance of the GUARDS architecture.
- Perform an early validation of the timing requirements of the resulting software system.

The ADE uses the generic architecture specification and the application requirements to provide the basic inputs to the validation and integration activities (see Figure 7.1).

D. Powell (ed.), A Generic Fault-Tolerant Architecture for Real-Time Dependable Systems, 121–138.

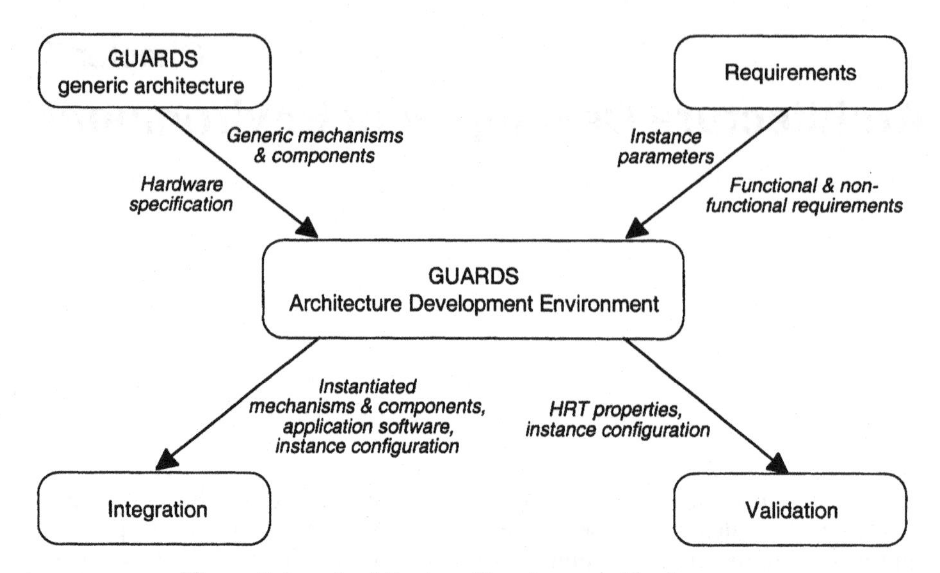

Figure 7.1 — Architecture Development Environment

The ADE provides the support for the design of the application software according to both functional and non-functional requirements. In particular, it takes into account the performance requirements of the application, as it builds a timing model of the whole system, based on a set of timing attributes associated with each hardware and software component.

This timing model is built with the Temporal Properties Analysis Tool-set (which is an integral part of the ADE itself) in order to verify that relevant hard real-time constraints (such as periods, budget times and deadlines) are respected.

The GUARDS methodology for the development of a software application is described in the next section. The following sections explain three main ADE design activities: Functional Architecture (Section 0), Infrastructure Architecture design (Section 7.3) and Physical Architecture design (Section 7.4). The ADE design activities are supported by a set of appropriate tools, which are described in Section 7.5.

7.1 Development Approach

The design and development of a GUARDS software application is based on a hard real-time (HRT) design method, which allows real-time requirements to be taken into account and verified during the software design. The method also addresses the problem of designing replicated fault-tolerant architectures, where a number of computing and communication boards interact for the consolidation of input values and output results.

The design of a GUARDS application is defined as a sequence of correlated activities, that may be re-iterated to produce a software design that complies with both the functional and non-functional requirements of the application. Three design activities are identified:

- *Functional Architecture design*, where the software application is defined through an appropriate design method and according to its functional requirements and its performance requirements (task periods, deadlines, etc.).
- *Infrastructure Architecture design*, where the required hardware boards and generic GUARDS software components are identified. They constitute the underlying computing environment of the application software.
- *Physical Architecture design*, where the Functional Architecture is mapped onto the Infrastructure Architecture and analysed according to the performance requirements. This is done not only for the processors within each replicated channel, but also at the inter-channel level, to determine the ICN exchanges needed to consolidate input values and output results.

The Functional Architecture embodies commitments that can be made independently of the constraints imposed by the execution environment, and is primarily aimed at satisfying the functional requirements (although the existence of timing requirements, such as end-to-end deadlines, will strongly influence the decomposition of the Functional Architecture).

The Physical Architecture takes these functional requirements and other constraints into account, and embraces the non-functional requirements. The Physical Architecture forms the basis for asserting that the application's non-functional requirements will be met once the detailed design and implementation have taken place. It addresses timing requirements, and the necessary schedulability analysis that will ensure (guarantee) that once built, the system will function correctly in both the value and time domains (within some failure hypotheses).

The Functional Architecture must be mapped on top of an instance of the GUARDS generic architecture (Figure 7.2). A further step in the architectural design is thus needed in order to define the components of the instance in terms of the dependability requirements: a number of channels, node configuration in the channels, type and version of mechanisms and components to be included in the executable code running on the nodes, and connecting hardware (ICN and channel communication nodes). This design activity is called Infrastructure Architecture design.

The Physical Architecture activity has to take into account the output of both Functional and Infrastructure Architecture. During the Physical Architecture activity it is possible to verify the temporal properties of the software system; the results of the analysis may have a feedback on all architectural activities. In addition, further steps of software development may cause the architecture to be revisited. The impact of the timing analysis on the architecture is the rational behind the choice to include the tools dedicated to such analysis in the ADE.

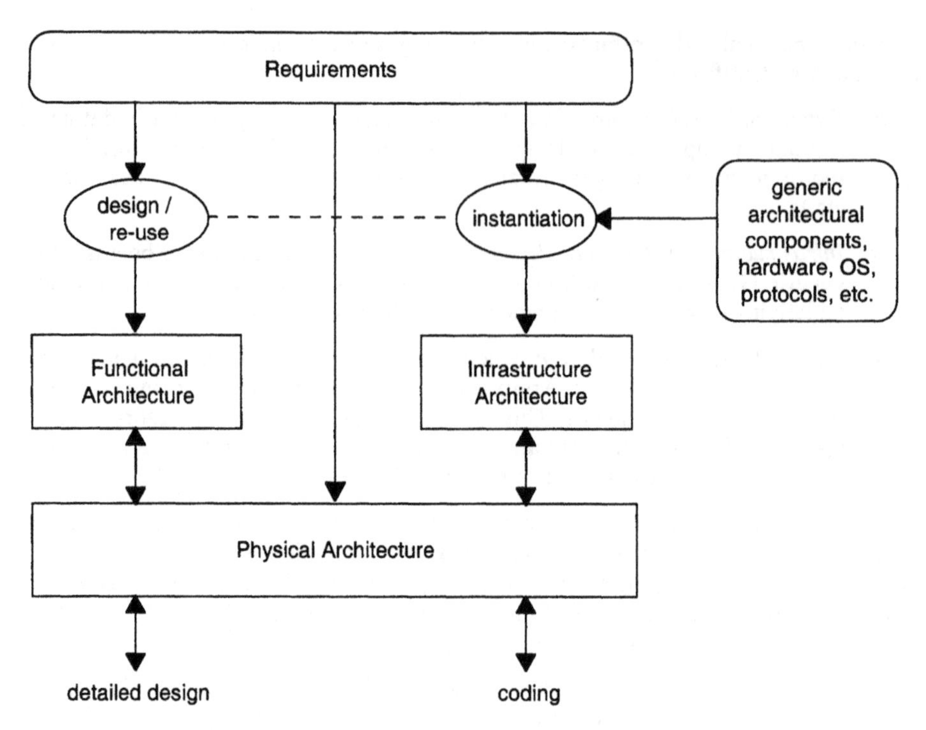

Figure 7.2 — ADE Design Activities

7.2 Functional Architecture Design

This activity is mainly driven by the functional requirements and is as independent as possible from the details of the GUARDS instance. This allows the design of software systems suitable to be reused (totally or partially) with different instances of the GUARDS architecture. Nevertheless, the choices made during this activity can heavily affect the mapping of the resulting architecture onto the result of the Infrastructure Architecture. The designer must be aware, at least, of the consequences of high level choices on the integration of the actually instantiated mechanisms.

The results of further activities (Physical Architecture in particular) will generally lead to refinements of the Functional Architecture to optimise the interactions with the underlying platforms.

7.2.1 Choice of Design Method and Tool

The design of the Functional Architecture is supported by an appropriate method and tool. To provide for genericity, GUARDS does not force the selection of a specific method, but it is assumed that the method selected by the user is suitable for the design of real-time software systems. Nevertheless, a survey and analysis of design methods has shown that only HRT-HOOD [Burns & Wellings 1994] addresses explicitly the design of hard real-time systems, providing means for the verification of their performance. The use of a hard real-time design method gives warranties about the ability to build a software system suitable for timing analysis. Therefore, HRT-HOOD has been selected as the baseline design method within GUARDS and HRT-HoodNICE (by INTECS Sistemi) adopted as supporting tool [Intecs-Sistemi 1996].

However, the same analysis revealed a few weaknesses of the method, in particular related to the design of distributed systems. Improvements to the method have been defined (with the contribution of the same authors of HRT-HOOD from the University of York), embodying in it a concept of "Virtual Nodes" similar to the one present in the HOOD 3.1 method [ESA 1991]. The new concept allows the method to take into account the lane dimension of GUARDS (by allocating objects to different nodes within a channel) and the integrity levels (by defining "spatial firewalls" around objects of a given criticality). The HRT-HoodNICE Tool-set has been enhanced accordingly.

The use of HRT-HoodNICE is not however compulsory for the GUARDS end-users. The choice of a design method and the related tools depends on a number of constraints, such as the customer preferences, the end-users' current practice and the application domain standards. The ADE user is free to use the preferred software design tool but has to be prepared for some consequences:

- General rules and constraints applicable for the development of hard real-time software, aimed at obtaining predictable systems, have to be followed.
- A number of automatic actions that support the overall software development process are tightly linked to the use of HRT-HoodNICE. Such utilities are not available if a different design tool is used. In order to minimise the set of lost capabilities, the other GUARDS ADE tools are made as independent as possible from the design tool, and precise interfaces are defined.
- The enhancements of HRT-HoodNICE, made for the GUARDS project, minimise the amount of information that has to be entered in the tool in order to use the tool's most important features. The end-user can thus create an HRT-HOOD "image" of the application's Functional Architecture with a limited effort.

7.2.2 HRT-HOOD

Hard Real-Time Hierarchical Object-Oriented Design (HRT-HOOD) [Burns & Wellings 1995b] is a structured design method based on the HOOD design method [Robinson 1992] and adapted for use in hard real-time systems. The most important stage in the development of any real-time system is the generation of a consistent design that satisfies a specification of requirements. However, real-time systems differ from the traditional data processing systems in that they are constrained by certain non-functional requirements. Non-functional requirements include dependability (e.g., reliability, safety), timeliness (e.g., responsiveness, temporal predictability), and dynamic change management. It has been common practice for system developers, and the methods they use, to concentrate primarily on functionality and to consider non-functional requirements comparatively late in the development process.

Typically standard design methods do not have adequate provisions for expressing these types of constraints. HRT-HOOD provides a framework within which the properties of real-time applications can be expressed. It gives the explicit recognition of the types of activities that are found in hard real-time systems (i.e., cyclic and sporadic activities). In addition, the resulting design is amenable to schedulability analysis.

Originally, HRT-HOOD was targeted at Ada system development and provides strict guidelines on the systematic translation of designs to Ada code. These guidelines and computational model are embodied in a CASE tool environment HRT-HoodNICE [Intecs-Sistemi 1996]. The computational model and scheduling model embraced by HRT-HOOD is an event and time-driven computational model within a pre-emptive priority-based scheduling framework.

Any hard real-time design method must constrain the design process if it is to produce analysable software. HRT-HOOD is explicitly designed to ensure that system decomposition conforms to a set of constraints that facilitates analysis of the final system.

The Functional Architecture design activity ensures that the design conforms to a computational model which facilitates timing analysis. The outcome of the Functional Architecture design activity is a collection of terminal objects (objects which do not require further decomposition) with all their interactions fully defined. Although this decomposition is essentially functional, the existence of timing requirements is a major driver in the construction of the application's architecture. The exact form this analysis must take is not defined by HRT-HOOD. However, the HRT-HOOD design process has been successfully integrated with the use of static priority analysis and pre-emptive dispatching.

The Physical Architecture addresses the non-functional (e.g., timing) requirements and the constraints of the execution environment. The activities undertaken during the Physical Architecture phase define priorities, offsets and (timing) error conditions that will be accommodated. The result is that timing behaviour is guaranteed.

HRT-HOOD attempts to be independent of any scheduling theory that might be used to guarantee the timing properties of programs. Instead, it provides a framework within which the properties of real time applications can be expressed. Once the Physical Architecture design phase is complete, detailed design and coding can procede. The key issue here is to constrain the coding style to one that is amenable to (non-pessimistic) worst-case execution time analysis.

HRT-HOOD was not developed from scratch. The two basic object types from the HOOD method — passive and active objects — are supplemented in HRT-HOOD by protected, cyclic and sporadic objects, thus leading to a total of five object types:

- **Passive** objects have no control over when invocations of their operations are executed, and do not spontaneously invoke operations in other objects.

- **Active** objects may control when invocations of their operations are executed, and may spontaneously invoke operations in other objects. Active objects are the most general class of objects and have no restrictions placed on them.
- **Protected** objects may control when invocations of their operations are executed, and do not spontaneously invoke operations in other objects; in general protected objects may not have arbitrary synchronisation constraints and must be analysable for the blocking times they impose on their callers.
- **Cyclic** objects represent periodic activities, they may spontaneously invoke operations in other objects, but the only operations they have are requests that demand immediate attention (asynchronous transfer of control (ATC) requests).
- **Sporadic** objects represent sporadic activities; sporadic objects may spontaneously invoke operations in other objects; each sporadic has a single operation which is called to invoke the sporadic, and one or more operations which are requests which demand immediate attention (they represent asynchronous transfer of control requests).

HRT-HOOD enhances HOOD to explicitly address the needs of hard real-time system design. The enhancements include:

- Explicit recognition of the types of the typical activities (i.e., cyclic and sporadic) of hard real-time systems.
- Integration of appropriate scheduling paradigms with the design process.
- Explicit definition of the application timing requirements for each activity.
- Definition of the relative importance (criticality) of each activity.
- Decomposition to a software architecture that easily allow the processor allocation, schedulability and timing analysis.
- Facilities and tools to allow static verification as early as possible.

HRT-HOOD extends the basic HOOD Active object type to specify cyclic and sporadic activities and protected resources (HRT objects), and defines a set of rules for their decomposition and interaction. HRT objects are annotated with a set of attributes related to their timing constraints (e.g., periods, worst-case execution times, deadlines, etc.), which are used to derive a timing model suitable for being analysed according to the Deadline Monotonic Scheduling algorithm and the Immediate Priority Ceiling Inheritance protocol (for more details, see Chapter 3).

The design process starts with a single object, called the root object. This is decomposed into a set of child objects. The decomposition process continues until all the objects in the design that need to be analysed for scheduling are single threaded. This level of the design is called the terminal level. A hard real-time system designed using HRT-HOOD will contain at the terminal level only cyclic, sporadic, protected and passive objects. Active objects, because they cannot be fully analysed, will only be allowed for background activity. The computational model that underlies HRT-HOOD consists of a number of active, cyclic and sporadic concurrently executing threads communicating with each other (more details on the GUARDS computational model can be found in Chapter 3).

7.3 Infrastructure Architecture Design

The Infrastructure Architecture design activity is supported by the Infrastructure Design Tool (Figure 7.3) managing an archive of generic hardware and software components which can be navigated through in order to select the elements needed to define specific GUARDS instances. Such components are described by their relations and compatibilities, and by the relevant performance attributes.

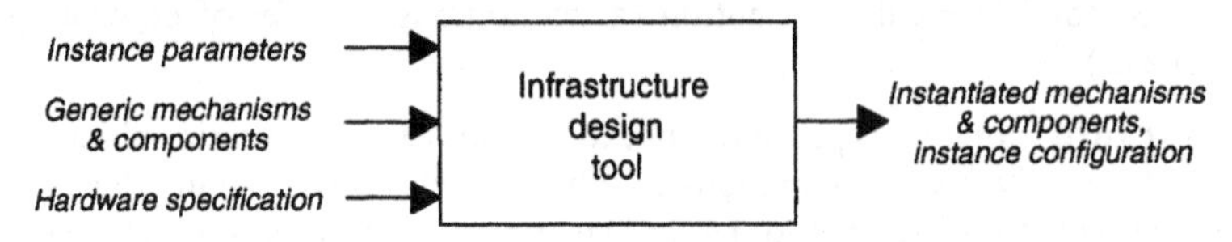

Figure 7.3 — Infrastructure Design Tool

In the following sections the specific interfaces of the Infrastructure Design Tool to the other development activities are detailed.

7.3.1 Instance Parameters

Instance parameters are the result of a global assessment of the application requirements, in particular the dependability requirements. Inputs to the Infrastructure Design Tool are the properties of the desired GUARDS instance in terms of hardware/software configuration. The following list is an incomplete specification, which is refined during the detailed design and implementation phases:

- Number of channels, C
- Number of lanes, L
- Number of integrity levels, I
- Application implementation language
- Mechanisms needed and their required dependability level

This input is contained in a text file, in order to make the tool-set independent from the overall configuration of the development environment.

7.3.2 Generic Mechanisms and Components

Generic mechanisms and components are held in an archive. Each mechanism corresponds to a given dependability property or feature that can be conferred on a GUARDS instance. A mechanism is composed of GUARDS component(s) which implement the dynamic part of the mechanism, and methodological rules that describe the design rules that must be followed during the development, in order to ensure a correct implementation of the mechanism.

For each mechanism the archive contains one or more sets of components. Different versions of the mechanism implementation may be maintained in the archive. The different versions of a given mechanism generally differ in:

- Supported set of C, L, I values
- Level of dependability
- Reconfiguration capabilities

For each set of components the archive contains:

- Data concerning the code to be run to implement the mechanism. Executable and object files are provided, grouped with respect to the different types of targets. A classification of targets is: i) the ICN hardware; ii) nodes dedicated to interface a channel to the ICN; iii) other nodes in the channels.
- Data (or references) concerning the hardware platforms (for each category) suitable to support the mechanism implementation.
- Data concerning the dependability properties of the particular version of the mechanism. This information is used by the Infrastructure Design Tool to select the configuration(s) of the mechanism that match the dependability level specified by the GUARDS instance parameters.
- Information relevant to the validation activities.
- Data concerning the inter-operability of the concerned mechanism with other mechanisms and with the application software.
- The specification of the methodological rules to be followed in the application design.

7.3.3 Hardware Specification

Hardware specification is an archive containing a description of each hardware component suitable to be included in an instance of the GUARDS architecture. Components are either GUARDS hardware components or selected COTS hardware components. For each component a substantial amount of information of various kinds is to be maintained for general project purposes. In the following, only the data directly concerning the Infrastructure Design Tool are mentioned:

- Data concerning the dependability properties. This information is used by the Infrastructure Design Tool to select the configuration(s) of the hardware components that match the dependability level specified by the GUARDS instance parameters.
- Data concerning the inter-operability of the concerned component with the other components of the GUARDS architecture instance.
- Data concerning the hardware/software interfaces, mainly references to technical documentation.
- Data concerning the hard real-time properties of the component for the purposes of schedulability analysis. It has to be noted that the values

contained in the Platform Data File describe the performance of the platform defined by the hardware and the actual operating system.

The output of Infrastructure Design Tool is GUARDS Instance Configuration stored in a text file. A preliminary list of its contents is given below:

- A description of the instance's hardware configuration. Each hardware component is identified by a unique name.
- The list of the mechanisms implemented in the instance and the related hardware and software components.
- The methodological rules and the inter-operability instructions extracted from the software components archive for each mechanism. They are both an input and an obligation for the application software design activity.
- For each hardware component, the relevant description extracted from the hardware components archive.
- The software configuration in terms of original executive and GUARDS software components implementing the mechanisms.
- Information and guidelines related to the validation activity as specified in the archive of software components.

7.3.4 Instantiated Mechanisms and Components

The main activities related to the Infrastructure Architecture design process are the following (see Figure 7.4). First, a search is made into the software and hardware archives in order to find one or more configurations matching the specified GUARDS Instance Parameters.

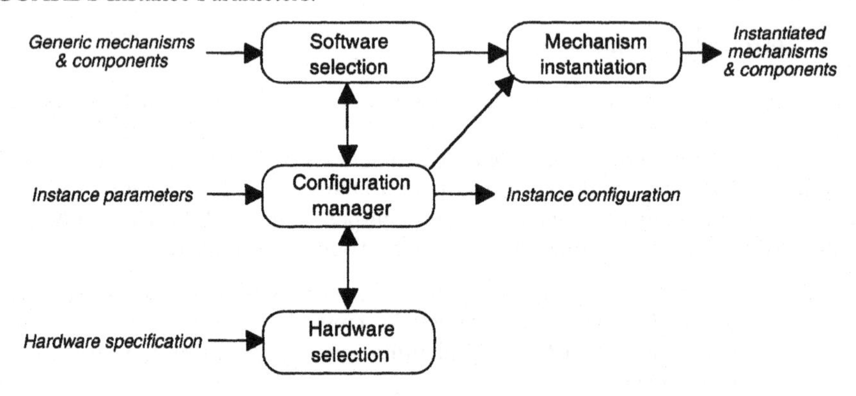

Figure 7.4 — Infrastructure Design Activities

The user is expected to interact with the tools in order to resolve multiple possibilities. The Configuration Manager is in charge of checking the feasibility, inter-operability and suitability of a proposed configuration.

After the elaboration of the GUARDS Instance Configuration it is possible to instantiate the mechanism in the desired form. The Mechanisms Instantiation can be a simple copy of software modules from the archive, but can also include some elaboration of a generic implementation in order to adapt it to the specific environment. The instantiation of a mechanism might also require the compilation of data tables to be used by the software components. In practice, there is no limit to the automatic instantiation capabilities once appropriate conventions are established for the mechanisms, components and application software development.

7.4 Physical Architecture Design

Physical Architecture design activity defines the replication/distribution of the Functional Architecture elements over the GUARDS architecture instance defined by the Infrastructure Architecture.

The purpose of this activity is threefold:

- To identify the whole set of software elements to be loaded on each node (both application software and mechanisms/components software).
- To ensure that these software elements are correctly interfaced.
- To check that they can run together without violating the timing requirements.

The ADE supports the Physical Architecture activity by means of HRT-HoodNICE, the Infrastructure Design Tool and the Temporal Properties Analysis Tool-set.

As part of the Physical Architecture activity, the application tasks (i.e., the HRT objects) identified in the Functional Architecture are mapped on the Infrastructure Architecture and coupled with the real-time models of the selected components, in order to analyse and verify their schedulability properties. This is done by the Temporal Properties Analysis Tool-set, which analyses the performance of the resulting distributed software system.

The tool-set includes a Schedulability Analyser and a Scheduler Simulator, based on those available in HRT-HoodNICE. They have been properly enhanced in order to provide a more precise and realistic analysis (by taking into account the concept of "offsets" in the activation of threads, see Chapter 3, Section 3.2.10), and to cope with the specific needs of a redundant fault-tolerant architecture (by allowing the analysis of the interactions over the inter-channel communication network).

A further result of the Physical Architecture is that, on the basis of the real-time models produced by the Schedulability Analyser, the critical interactions among software functions on different channels are scheduled in a deterministic way and the ICN transfer slots allocated to them.

According to the dependability requirements, each critical application task replica needs to consolidate its inputs and output results with those of the corresponding replicas on the other channels (Figure 7.5).

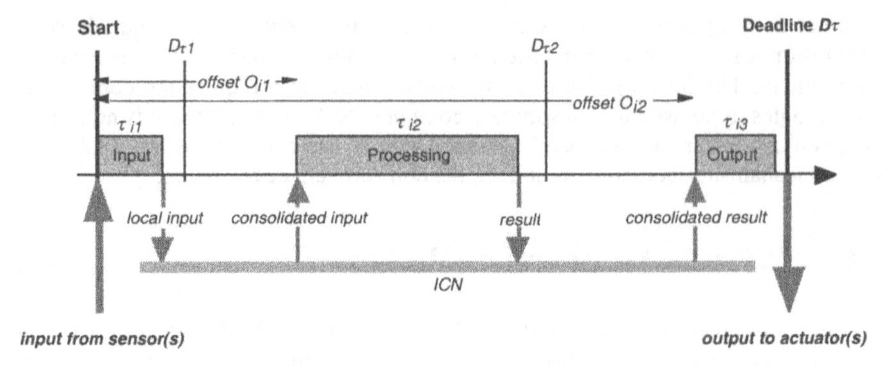

Figure 7.5 — Decomposition of Critical Tasks

Each application task is structured as real-time transaction, consisting of three sub-tasks, or threads:

1. The input value is acquired on each channel;
   ⇒ the "local" input values are consolidated over the ICN.
2. The output results are calculated on each channel;
   ⇒ the results are consolidated over the ICN.
3. The consolidated result is actually produced;
   ⇒ the consolidated result is output to the actuator(s).

For each task τ, a deadline D_τ is set that defines the time by which the final value must be sent to the actuator(s) (corresponding to the third thread of the transaction). Intermediate deadlines $D_{\tau 1}$ and $D_{\tau 2}$ are also introduced for the first and second threads. They define the time by which the input or output results are (or must be) ready for transfer through the ICN (after a fixed intra-channel transfer time) and consolidated. The transfer and consolidation of each value over the ICN must take place at pre-defined transfer slots (to synchronise such activities on each channel) and the needed duration determines an "offset" for the activation of the following thread (Figure 7.6).

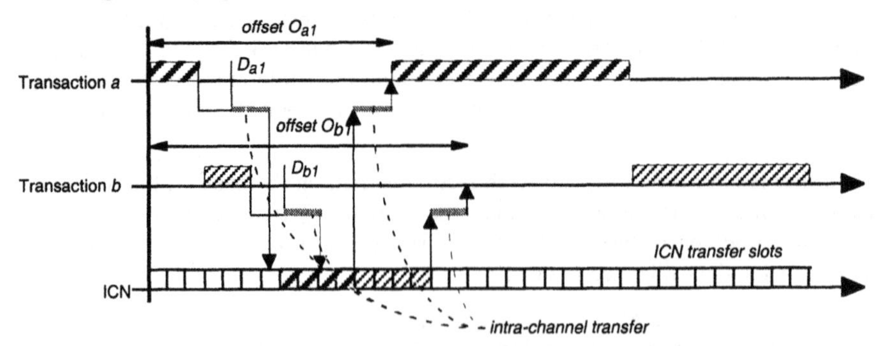

Figure 7.6 — Inter-Channel Schedulability

Although the final deadline is set by the requirements, it should be noted that the intermediate deadlines can be set arbitrarily during the design, according to the intra-channel schedulability analysis and the allocation of ICN transfer slots; changing them may imply different ICN transfers (i.e., slot allocations) and different offsets. Consequently, the HRT schedulability analysis (at the intra-channel level) must take into account the possible tuning of HRT design attributes (i.e., intermediate deadlines and offsets), as well as the slot allocation (i.e., the inter-channel schedulability).

7.5 Tool Support

The tool support for ADE is shown in Table 7.1. The Functional Architecture activity is supported by HRT-HoodNICE or by any suitable software design tool selected by the end-user. The Infrastructure Design Tool is a tool specifically designed and implemented for the GUARDS project to support the Infrastructure Architecture activity during which the hardware and generic software components are identified. Once the application tasks identified in the Functional Architecture are mapped on the selected components, the tools that support Physical Architecture activity analyse and verify their schedulability properties. A description of the tools will be given in the following sections.

Table 7.1 — ADE Tools

Design Activity	Tools
Functional Architecture	HRT-HoodNICE
Infrastructure Architecture	Infrastructure Architecture Tool
Physical Architecture	Schedulability Analyser Simulator Tool ICN Slots Allocator

7.5.1 HRT-HoodNICE

HRT-HoodNICE [Intecs-Sistemi 1996] is an extension of the HoodNICE tool-set intended to support the HRT-HOOD method version 2.0 [Burns & Wellings 1993]. The extensions include the database definition, the graphical editor, the textual editors, the off-line tools (such as the Rule Checker and the Document Generator) and the Code Extractor.

HoodNICE [Intecs-Sistemi 1997] is a tool-set, developed by Intecs Sistemi SpA, which supports the architectural and detailed design activities relating to the production of Ada, C and C^{++} software. The main features of HoodNICE are:

- Full support and conformance to the Version 3.1 of the HOOD method [ESA 1991]
- Support for the derivation of Ada, C and C++ code from a HOOD design
- Advanced user interface techniques, such as graphics, direct manipulation, multi-windows and menus

- Support for multi-user access to the project database
- Configurable to interface third party tools
- High degree of configurability by the user
- Import/export facilities in Standard Interface Format (SIF)
- Optional interfaces with a range of documentation systems (notably TPS, Framemaker, DecWrite, Postscript and LaTex)

HoodNICE is designed to be portable across different software and hardware platforms that provide OSF/Motif (version 1.2). The tool-set architecture is shown in Figure 7.7. In addition to the component tools, it consists of:

- The **Dialog Manager**, which handles the interactions with the user providing the interactive tools with the presentation and dialog services.
- The **Data Handler**, which manages all the accesses to the Data Base, making transparent to the tools its implementation on different platforms.
- The **Integration Manager**, which provides the tools with integration facilities and consistency check services.
- The **External Tools Manager**, which is a special tool for managing interactions with external tools, namely tools that are executed as separated processes and that do not use the Data Handler to access the database. HoodNICE provides mechanisms for activating external tools and for interacting with them through a Remote Procedure Call protocol.

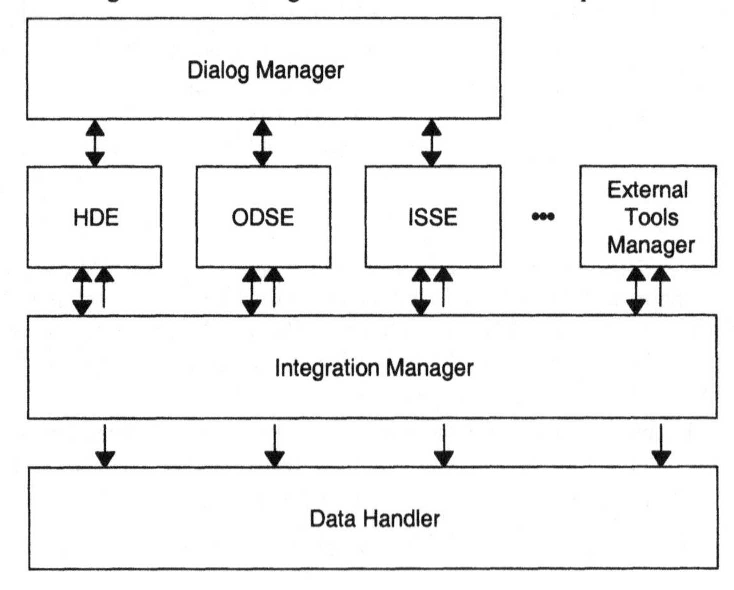

Figure 7.7 — HoodNICE Tool-set

The component tools are:

- The **HOOD Diagram Editor** (HDE) is a graphic multi-window syntax-driven editor supporting the specification of a HOOD objects, relationships among them, and their hierarchical decomposition.
- The **ODS Editor** (ODSE) supports the specification of objects using the textual formalism of the HOOD method, the ODS (Object Definition Skeleton). It is a template driven editor that allows the edited text to be compiled on line, checking its syntactic and semantic correctness.
- The **Informal Solution Strategy Editor and Analyser** (ISSE) provides a text editor to edit the Informal Solution Strategy text and an Analyser to analyse the text in order to identify the objects and the operations that formalise the informal solution.
- The **Text Attribute Editor** is a general purpose free text editor for editing textual attributes associated with objects.
- The **Document Generator** enables automatic generation of design documentation by extracting and formatting the information contained in the project database. The document generation is driven by a template defined by the user.
- The **Code Extractor** allows Ada or C code to be automatically extracted from a HOOD design. For each object, the code extractor generates a set of source files, and for the whole design the main unit of the program.
- The **SIF Generator and Compiler** are provided for generating the SIF (Standard Interchange Format) representation of a HOOD design (or part of it) and for compiling a SIF file generating the HoodNICE internal representation.
- The **Rule Checker** analyses the design and generates a report in which all the violations of the rules are outlined. The user can select the objects on which to apply the check, and the rules that have to be checked.

The HRT-HOOD extensions to the basic HoodNICE Tool-set include:

- Definition of HRT object and operation attributes in the database — object types are: passive, active, protected, cyclic, sporadic; attributes include: priorities, periods, offsets, deadlines, etc.
- Definition of HRT operation constraints in the database — protected operations and asynchronous transfer of control requests are supported.
- HRT attributes inheritance and feedback — when an object is decomposed its applicable HRT attributes are automatically inherited by the child objects depending on their type. A feedback from child to parent HRT attributes is also automatically executed after changes to (a subset of) the HRT attributes.
- Enhanced Hood Diagram Editor features — specific commands are implemented to create HRT objects, activate the Temporal Properties Analysis Tools and display their results.

- ODS and SIF extensions — both ODS and SIF are extended with a section to define the HRT attributes.
- Off-line Checker extension — the Off-line Checker of HoodNICE is extended to enforce the HRT-HOOD rules concerning the "use", "include" and "implemented by" relationships. It is also able to detect inconsistencies between the HRT attributes of a non-terminal object and its child objects.
- Code Extractor Enhancement — the Code Extractor of HoodNICE has been enhanced to generate the skeleton of the whole system. The automatic extraction includes all the aspects related to the implementation of threads and their interactions in terms of synchronisation and communication. The user is then in charge of supplying the functional part of the system, that is the sequential code of the operations' body.

Inputs to the Temporal Properties Analysis Tool-set are the relevant figures of the used run-time system (Ada run-time support or an operating system kernel) on the specific target, and the description of the system in a custom language. Such figures are currently contained in a couple of files, the Platform File and the Timing Data File. The Platform File contains the relevant performance figures of the adopted run-time system on the specific target. They include, for example, the context switch time, the resources' enter and leave times, the maximum overhead of run-time primitive and others. The Timing Data File contains the model of the software system, automatically derived from the HRT objects.

7.5.2 Infrastructure Architecture Tool

The Infrastructure Architecture Tool (also known as the GUARDS Configuration Manager) is a database manager intended to manage the complexity of the configuration of a GUARDS architecture instance from a number of existing software and hardware components. The generic GUARDS architecture defines a number of mechanisms, each of which is composed of several hardware and software components. They are stored in the archives of mechanisms and components. A mechanism is composed of a number of components and a given component can be part of several mechanisms. A GUARDS generic architecture instance implements some of (or all) the defined mechanisms. An archive is used to store data concerning instances.

7.5.3 Schedulability Analyser

The Schedulability Analyser is in charge of verifying that the system is schedulable in its worst case of execution given the interactions among the concurrent threads and the execution time of the sequential pieces of code. Inputs to the tools are the relevant figures of the Ada run-time support or operating system services on the specific target, and the description of the software system in a custom language. Such data are currently contained in the Platform Data File and the Timing Data File. During the execution of the Analyser, the user is requested to provide some inputs. The input choices are related to the selection of the blocking approach (e.g., Disable Interrupts) and cyclic object transformations (if requested). The results of the analysis are contained in two output files:

- A summary of the analysis, containing the computed priority and the schedulable/un-schedulable information for each thread and, when schedulable, the time at which the individual thread meets its deadline.
- A copy of the Timing Data File annotated with the computed priorities. This file is also one of the inputs of the Scheduler Simulator tool.

7.5.4 Scheduler Simulator

The Scheduler Simulator is in charge of performing a run-time simulation of the software system behaviour in terms of predicted scheduling events and their associated time of occurrence. The Scheduler Simulator also computes the number of activations for each entity and the CPU load. The simulation is not necessarily executed in the worst execution case. Therefore, it may happen that entities flagged as UNSCHEDULABLE by the Analyser may execute to completion during the simulation.

7.5.5 ICN Slot Allocator

The ICN Slots Allocator (called "GASAT" — GUARDS Allocator Schedulability Analyser Tool) takes as its input the timing model of a GUARDS application and analyses it in order to allocate the slots related to each communication on the ICN Table needed by the application tasks. It checks and sets the offset values of each thread and consolidates the schedulability analysis at inter-channel level.

The ICN Table is divided into cycles and each cycle into slots. The user can specify the initial ICN Table (for instance to specify slots reserved for special needs, like ICN synchronisation) by setting the number of cycles and the number of slots per cycle and filling the slots. The input file contains a description of the application being analysed in terms of transactions, threads and shared data. All the threads belonging to a transaction are supposed to be executed sequentially. At the end of its execution, a thread may require an inter-channel data exchange to perform a consolidation of its values (a vote or an interactive consistency exchange). In this case, based on the transaction's period and deadline, the tool reserves a number of adjacent slots to perform the requested operation. It also checks the offset of the next thread in the transaction, and shifts it (if needed) until it matches the performed slot allocation. When all transactions are analysed and the inter-channel schedulability is consolidated, the tool produces the ICN Table which supports the ICN Manager to drive its operations.

7.5.6 Use of the Tools

The relations and dependencies between the tools constituting the GUARDS ADE are shown in Figure 7.8. The software application is first designed using the HRT-HOOD method (and HRT-HoodNICE tool), identifying in particular the HRT objects (that is, "Cyclic", "Sporadic" and "Protected" objects). The control flows of these objects are then implemented as independent "threads". Real-time properties like periods, execution times and deadlines (the HRT model) are also associated to them according to the related requirements and constraints. According to the dependability requirements, the overall system architecture (the "infrastructure") is defined in terms of number of computers, communication media, operating systems and fault-tolerance mechanisms. The performance characteristics of the selected hardware and software components are then extracted.

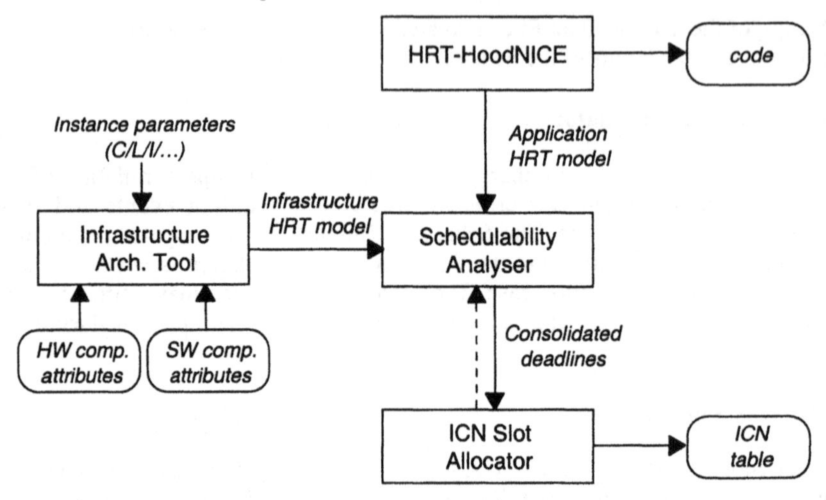

Figure 7.8 — Relationship between Tools

The real-time attributes are extracted from the produced software design and merged with the performance characteristics of the system architecture. They represent the real-time model of the whole system. The temporal properties of the model are analysed according to stated performance requirements in order to verify that all stated deadlines are satisfied and that the software has the expected run-time behaviour. The analysis takes into account the requirements for inter-channel interactions, allocating the slots for data transfers and interactive consistencies over the ICN. As a result, a number of data tables are consistently generated. Such tables will drive the ICN functions that, due to their criticality, have to be executed deterministically.

If the requirements are not satisfied, the designer can modify the design or select a hardware or software component with better performance, and re-verify the new model in an interactive and consistent manner. Once the performance requirements are satisfied, the design can be mapped into code, according to the characteristics of the designed "active" objects and their relations. The overall structure of the software application is extracted from the HRT-HOOD design and the related code is automatically generated. The generated code implements all the dynamic properties of the HRT objects, including their interactions. It represents the "skeleton" of the system, which is completed with the code implementing the functional requirements (produced manually or through other means).

Now the software system can be built, linking together the automatically generated code, the manually implemented functions and the produced data tables. It can be executed and tested. Performance measures can also be taken and the temporal property analysis consolidated. Should the need for some design modification arise, the relevant steps in the process can be re-iterated, in a simple, consistent and cost-effective way.

Chapter 8

Formal Verification

Formal approaches have been used in GUARDS as one of the elements of the validation strategy. In particular, formal verification has concentrated on four dependability mechanisms, namely: a) clock synchronisation, b) interactive consistency, c) fault diagnosis, and d) multi-level integrity. The first three mechanisms constitute basic building-blocks of the architecture, and the fourth one corresponds to a major innovation. The formal approaches applied include both *theorem-proving* and *model-checking*.

The work on the verification of clock synchronisation used theorem-proving and relied heavily on PVS (*Prototype Verification System*) [Owre *et al.* 1996]. Agreement and accuracy properties were proved on the high order logic specification of the mechanism. It led to the development of a general theory for averaging and non-averaging synchronisation algorithms [Schwier & von Henke 1997]. The verification of the synchronisation solution used in GUARDS was derived as an instantiation of this general theory [Canver *et al.* 1997].

The verifications concerning interactive consistency [Bernardeschi *et al.* 1997, Bernardeschi *et al.* 1998a, Bernardeschi *et al.* 1998c, Bernardeschi *et al.* 1999], fault diagnosis [Bernardeschi *et al.* 1998b] and multi-level integrity [Semini 1998, Fantechi *et al.* 1999] were all based on model-checking using the JACK (*Just Another Concurrency Kit*) tool-set [Bouali *et al.* 1994]. This integrated environment provides a set of specification and verification tools that were used to generate a state-based model of the behaviour of the mechanism under study and to prove properties formalised as temporal logic formulae: *Agreement* and *Validity* for interactive consistency, *Correctness* and *Completeness* for fault diagnosis, and the multi-level object *Segregation Policy* for multi-level integrity.

The adoption of model-checking as a formal verification technique was an innovative issue of the project: we will concentrate therefore on this technique, leaving to the end of the chapter a brief discussion of the relative merits of the techniques.

D. Powell (ed.), A Generic Fault-Tolerant Architecture for Real-Time Dependable Systems, 139–155.

8.1 Model-checking

Model-checking is an automated verification method for checking finite state systems against properties specified in temporal logic [Clarke *et al.* 1986]. The proof of the properties is carried out by means of exhaustive search on the complete behaviour (*model*) of the system.

The model of the system can be usually given by Kripke Structures (state-based approach) or by Labelled Transition Systems — LTSs — (action-based approach), which are essentially variants of classical state machines. They differ because in the former, states are labelled to describe how they are modified by the transitions, while in the latter, transitions are labelled to describe the actions which cause state changes. CTL [Emerson & Halpern 1986] is a temporal logic having Kripke Structures as interpretation domains. ACTL [De Nicola & Vaandrager 1990] (Action-based CTL) is a logic based on actions rather than states and thus is interpreted over LTSs. According to the selected model of computation, model-checking algorithms verify if a model satisfies a logic formula.

We adopted the action-based approach in GUARDS, since a specification in terms of LTSs appears more consistent with the original description of the mechanisms to be validated. Moreover, faults can be explicitly modelled by actions, thus allowing control over their occurrence. Failure modes are added to the specification and properties are proved on the model of the behaviour of the mechanism according to different fault hypotheses. Generally, the modelling of a mechanism is organised as follows:

1. Formal specification of the mechanism.
2. Injection of faults and failure modes in the specification, by extending the model to include the behaviour after faults; control of faults by using processes that constrain the actual occurrence of fault actions, according to some fault hypothesis [Bernardeschi *et al.* 1994].
3. Generation of the global model of the mechanism specified at step 1 and 2, as an LTS.
4. Use of the ACTL temporal logic to describe the properties that express the correctness of fault tolerance requirements on the mechanism, and automatic verification, by the JACK model checkers, of the ACTL properties on the LTS generated at step 3.

Since we follow the action-based approach to model checking, the LTS expressing the global behaviour of a mechanism has to be built. More formally, a mechanism is specified starting from a finite set of observable actions Act and giving the possible sequences of actions the mechanism can execute (mechanism's observable behaviour) by means of an LTS. We recall that an LTS is given by a set of states, and labelled transitions, which relate states:

Definition: An LTS is a 4-tuple $A = (S, s^0, Act_\tau, \rightarrow)$, where: S is a finite set of states; s^0 is the initial state; $Act_\tau = Act \cup \{\tau\}$ is a finite set of actions; and finally, $\rightarrow \subseteq S \times Act_\tau \times S$ is the transition relation between states. We denote by $s - a \rightarrow s'$, $a \in$

Act_τ, the transition from the state s to the state s' by executing a; in particular, s - a → s' indicates that a system in state s can perform a transition to state s' by executing the action a.

The special action τ, not belonging to Act, represents the unobservable action and it is used to model internal actions.

The specification of the mechanism is supported by the automatic tools available inside the JACK verification environment [Bouali *et al.* 1994]. Actually, a mechanism can be specified in JACK using two different techniques: textually by means of process algebra terms, or graphically by means of LTSs and networks of LTSs. A hierarchical approach to specification is possible: a single mechanism is obtained by composing the specification of its components.

8.1.1 CCS/Meije Process Algebra

The process algebra used in the project is called CCS/Meije. In CCS/Meije a system is described by a set of communicating processes. A process is defined as a set of states and transitions between states occurring by executing actions. Let Act be the set of observable actions; they are output actions if prefixed by the "!" character, or input actions if prefixed by "?". The process Q which executes action ?b followed by the action ?c and then stops is described as follows:

```
Q = let rec {
        S1 = ?b: S2
        and
        S2 = ?c: S3
        and
        S3 = stop
} in S1;
```

where the notation in S1 gives the initial state of Q.

Basic operators to compose processes are action prefix, choice, parallelism and restriction, as shown in Table 8.1. In particular, parallel composition of two processes corresponds to the interleaved execution of the two processes. Moreover, the two processes may synchronise on complementary input and output actions (i.e., actions with the same name but different prefix). The synchronisation is hidden by showing a τ internal action. The action restriction operator constrains processes to synchronise on the given actions.

The semantic models of the CCS/Meije process algebra are LTSs.

8.1.2 The Graphical Formalism

When a system is specified graphically, the LTS modelling the behaviour of the system is drawn or, if the system is composed by a set of subsystems, a *network* of LTSs is drawn. If a network is given, each LTS is surrounded by a box. The ports at the border of the box are the observable actions and can be used for synchronisation purposes. Moreover, a box can in its turn be specified as a network. If two boxes are drawn at the same level, they can synchronise via the actions they execute by linking the corresponding ports.

Table 8.1 — Some of the CCS/Meije Operators

Syntax		Meaning
a: P	Action prefix	Action *a* is performed and then process *P* is executed. The action may be prefixed by the *!* or *?* character
P + Q	Nondeterministic choice	Alternative choice between the behaviour of process *P* and that of process *Q*
P || Q	Parallel composition	The interleaved execution of process *P* and *Q*. The two processes may synchronise on complementary input and output actions, i.e. actions with the same name and different prefix
P \ a	Action restriction	Action *a* can only be performed synchronously, i.e. within a communication

The graphical formalism also provides the multiway synchronisation operator: more than two subsystems (boxes) synchronise when executing an action regardless of the prefix. Moreover, it is possible to make observable an action b on which subsystems synchronise by assigning a label to the link connecting the ports. In this case an action with the name of the label is shown when b is executed. Figure 8.1 shows the graphical specification of a network composed by two processes synchronising on action b. The initial state of an LTS is denoted by a double circle in the graphical formalism.

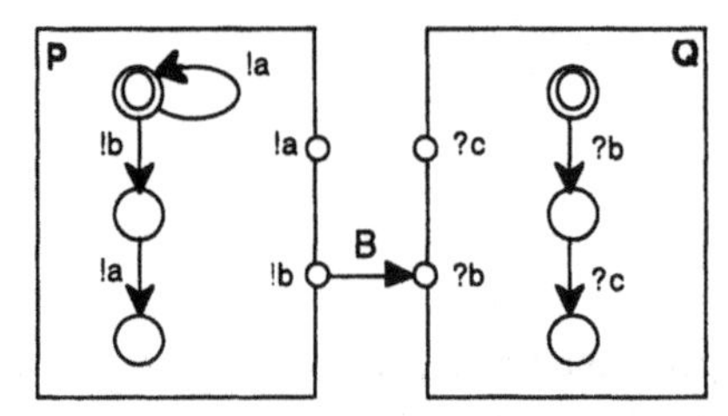

Figure 8.1 — A Network of Processes

8.2 Formal Specification of GUARDS Mechanisms

Each mechanism analysed in GUARDS can be considered as a distributed system that can be modelled as a set of synchronous processes in the considered formalisms. Since we assume the abstract view of a GUARDS mechanism as a distributed system, from now on we use the term *node* to generalise the concept of *channel* used for the GUARDS architecture. In the following, as an example, we present how the interactive consistency mechanism is modelled and verified. In particular, we present how the mechanism is formally specified, how possible failures are taken into account in the formal modelling, and how the properties of *Agreement* and *Validity* are proved on the global model of the behaviour of the mechanism in presence of faults according to given fault hypotheses.

8.2.1 Interactive Consistency Mechanism

Interactive consistency focuses on the problem of reliably distributing single source data to multiple nodes in presence of faults, see Chapter 2. The principal difficulty to be overcome in achieving interactive consistency is the possibility of arbitrary behaviour of faulty nodes: such a node may provide one value to a second node, but a different value to a third, thereby making difficult for the recipients to agree on a common value. Interactive consistency algorithms overcome this problem by using several rounds of messages exchange (phases in the following) during which node P tells node Q what value it received from node R and so on. In particular, the GUARDS interactive consistency mechanism implements a non recursive definition of the ZA algorithm described in [Gong *et al.* 1998], which is not in this case generic on the number of nodes and of rounds. In fact, the GUARDS architecture is designed to be exploited by three different pilot applications, namely a nuclear propulsion subsystem, a railway interlocking system, and a space-station payload control system. Of the three cases, the first employs an instance of the GUARDS architecture having just two nodes, where the interactive consistency mechanism does not apply, the second one employs three nodes and the latter four nodes. Therefore, only three and four node instances of the generic interactive consistency algorithm have been implemented. The implementation of the algorithm is organised in phases and is based on authenticated messages. Messages are saved into internal variables of each node and are checked to detect faulty nodes. In the last phase of the algorithm, a voting algorithm is executed by each node to reach agreement with the other nodes.

8.2.2 Specification of the 3-Node Interactive Consistency Mechanism

In the case of three nodes, the interactive consistency mechanism behaviour can be specified as a *network* of three processes P, Q and R, each executing the algorithm of a node. The processes communicate encoded values among them on dedicated links. The nodes synchronise at the beginning of each phase, by means of some common actions. We wish to note the following characteristics of the given specification of the interactive consistency mechanism:

1. Due to state explosion problems, we have to minimise the number of different values travelling on the links. Only 0's and 1's are considered as correct values.

2. Communication actions between nodes are seen as separate actions on separate links for each possible value transmitted. In this way we have non-parametric actions, that allow a simpler verification process. For example, node P sends to node Q the value 0 or 1 encoded by P, or by Q, or by R or the special value error. Moreover, the omission is modelled by an omission message. Then messages are modelled by the actions: !psendq_X where X = {encp_0, encq_0, encr_0, encp_1, encq_1, encr_1, omission, error}. Messages received by P from R are modelled by the actions: ?rsendp_X;

3. To maintain coherence with the original interactive consistency algorithm definition, we specify the variables as distinct processes inside a node, and the reading/writing of the variable are specified as actions performed by the related processes. For example, the textual specification of the variable p1 is

shown in Figure 8.2. Specifications of variables, such as this one, are automatically derived from the set of possible values.

```
varp1 = let rec {
  ZERO = !p1_eqto_0: ZERO +
             ?s_p1_1: ONE +
             ?s_p1_0: ZERO +
             ?s_p1_error: ERROR
and
  ONE = !p1_eqto_1: ONE +
             ?s_p1_1: ONE +
             ?s_p1_0: ZERO +
             ?s_p1_error: ERROR
and
  ERROR = !p1_eqto_error: ERROR +
             ?s_p1_1: ONE +
             ?s_p1_0: ZERO +
             ?s_p1_error: ERROR
} in ERROR;
```

Figure 8.2 — Specification of Variable p1

Phase 1 of the algorithm for node P can be given in a Pascal-like language as follows (cf. Chapter 2, Section 2.2.2.4):

```
vp:p:= p_encode(vp);
p_broadcast(vp:p);
//
msg1q:= q_receive();
msg1r:= r _receive();
```

Node P encodes its message vp and sends the message to other nodes. At the same time the node waits for the reception of the messages sent by the other nodes and saves them into two internal variables (msg1q and msg1r). This has been expressed in CCS-Meije as the process shown in Figure 8.3.

```
phase1 = let rec {
INITIAL =        !psend_vp_0: SEND_ZERO +
                 !psend_vp_1: SEND_ONE
and
SEND_ONE =       !psendq_encp_1: !psendr_encp_1: FIRST_REC
and
SEND_ZERO = !psendq_encp_0: !psendr_encp_0: FIRST_REC
and
FIRST_REC =      ?qsendp_encq_1: !s_x_encq_1: SEC_REC +
                 ?qsendp_encq_0: !s_x_encq_0: SEC_REC
and
SEC_REC =        ?rsendp_encr_1: !s_y_encr_1: END +
                 ?rsendp_encq_0: !s_y_encq_0: END
and
END = !startphase2: stop
}in INITIAL;
```

Figure 8.3 — Specification of phase1 of Node P

In the INITIAL state, the internal value vp of P is made observable to the external environment (!psend_vp_0 action or !psend_vp_1 action) and then either the state SEND_ZERO or the state SEND_ONE is reached. In such a state, the value of vp is encoded and sent to the other nodes (psendq_encp_0, psendr_encp_0 or psendq_encp_1, psendr_encp_1 actions); then a state is reached in which a message is received from node Q and saved into the variable X. After that, P reaches a state in which a message is received from node R and saved into the variable y. Finally, y synchronises with the other nodes to start the phase 2 of the mechanism (!startphase2 action).

Note that the parallelism between the broadcast of a message and the reception of messages from other nodes is realised as a sequence of actions. Similarly, the broadcast is realised as the sequential sending of messages to the participant nodes, thus choosing a possible implementation. We have formally proved that deadlock is avoided.

Figure 8.4 sketches the network of processes describing the behaviour of node P. Only some processes, ports and links are shown for readability. The complete specification of the node consists of 4 processes modelling the phases of the algorithm, 12 processes corresponding to the variables and 6 processes specifying the fault hypothesis (as explained in the next section).

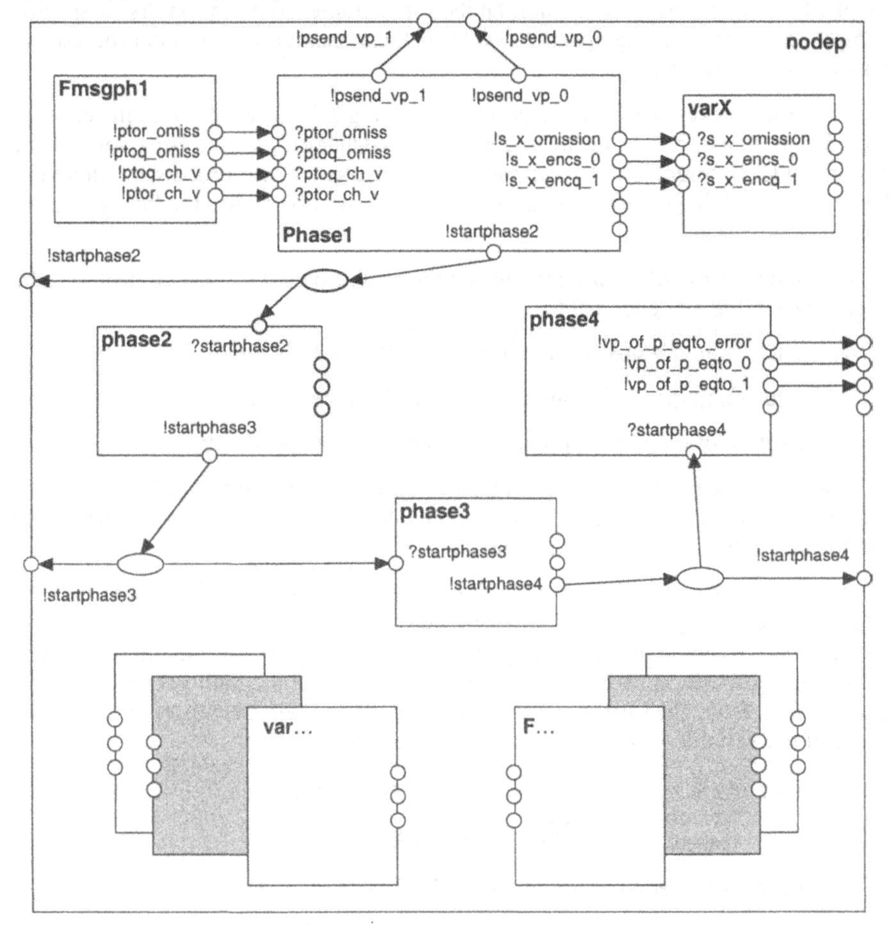

Figure 8.4 — Sketch of the Graphical Specification of Node P

8.3 Introducing Faults in the Specification

Starting from the specification representing the correct behaviour of a system (or sub-system), the specification of the system that may fail is obtained by introducing occurrences of the possible faults as actions and by adding the failing behaviour after the fault. New states and new actions, necessary to define the failure mode of the system, are generally introduced in the specification. If the action corresponding to a fault is executed, then the failure mode of the system is exhibited. Otherwise, the system goes on with its normal behaviour, maintaining the possibility to be disrupted by the fault action at the next step. The occurrences of faults are observable actions.

We can assume the more general view that any kind of fault may occur in any state and that the failing behaviour of the system depends on the point at which a fault occurs during the execution of the system. However, this generality can lead to models that are too large to be manageable. In specific cases, it is more convenient to take finer decisions on modelling faults; for example, confining faults to specific sub-processes or choosing specific points in the execution of the sub-processes at which a fault may occur.

Processes expressing the fault assumptions (single fault, arbitrary fault, etc.) are included as constraints on the behaviour of the mechanism; so, the behaviour of the system under different fault hypotheses can be studied. These processes, referred to hereafter as fault hypothesis processes, are composed in parallel with the system by means of synchronisation.

To summarise, three different techniques contribute to gain control on the occurrence of the modelled faults:

1. Inclusion in the specification of actions modelling faults.
2. Inclusion in the specification of the failure mode of the node.
3. Addition of the fault hypothesis processes to the specification.

The combined use of these techniques provides a great flexibility in the definition of the fault scenarios to be analysed. We illustrate the approach using the 3-node interactive consistency mechanism.

To analyse the introduction of fault occurrences in the specification of phase 1 of node P, let us consider the state FIRST_REC in which the message from Q must be received. A possible fault occurrence is the omission of the message. The following shows the specification of such state to include the fault occurrence. After an omission fault, the system reaches the state FIRST_REC_OMISSION in which the value of the variable X is set to omission (!s_x_omission action).

```
FIRST_REC = ?qsendp_encq_1: !s_x_encq_1: SEC_REC +
        ?qsendp_encq_0: !s_x_encq_0: SEC_REC +
        ?qsendp_omission: FIRST_REC_OMISSION
and
FIRST_REC_OMISSION = !s_x_omission: SEC_REC
```

In general, we do not need a new state for each occurrence of fault, thus optimising the specification. An example fault hypothesis process is the Fmsgph1 box in Figure 8.4, which controls the occurrence of faults when node P sends a message to the other nodes in the phase 1 of the algorithm.

The 3-node interactive consistency mechanism specification is composed of 69 processes. Some figures about the size of the generated models in terms of number of states of the LTSs are shown in Table 8.2.

Table 8.2 — Number of States of some Models Generated for the 3-Node Case

Model	Number of states
Node without faults	19714
Network of 3 nodes without faults	41863
Network of 3 nodes with an arbitrary faulty node	70120

8.4 Formal Verification

In the verification phase, the properties to be checked are expressed as ACTL formulae. The model checkers available inside JACK for ACTL are then used to show the validity of such properties on the mechanism.

8.4.1 The Logical Language

ACTL includes action formulae, state formulae and path formulae. An action formula expresses constraints on the actions that can be observed. In general, one or more actions may be possible in a given state of an LTS and each of these actions represents the beginning of an alternative continuation of the execution. A state formula describes the possible ways an execution can proceed. Finally, a path formula states properties of a single execution. The syntax of some basic ACTL operators and their informal meaning is shown in Table 8.3. Starting from these, some derived operators can be defined in the usual way. The formal semantic of ACTL is given over LTSs: informally, a formula is true on an LTS, if the sequence of actions of the LTS verifies what the formula states. The model checker AMC has a polynomial time complexity in terms of the number of states of the model and the length of the formula.

8.4.2 The 3-Node Interactive Consistency Mechanism

In order to validate the interactive consistency mechanism, *Agreement* and *Validity* properties have to be checked, see Chapter 2. The interactive consistency mechanism relies on given assumptions on the possible faults. In particular, the authentication assumption requires that messages sent by non-faulty nodes cannot be undetectably altered by relaying nodes. In the formal verification, we distinguish the following cases:

- No fault.
- Omission faults (phases 1, 2 or 3).
- Transmission faults leading to an incorrect signature (phases 1, 2 or 3).

- Value faults in phase 1: the faulty node sends a value different from the correct one (this may lead to inconsistent values received on remote nodes).
- Value faults in phases 2 or 3: the faulty node relays a modified message (these faults violate the authentication assumption).

Table 8.3 — Some of the ACTL Operators

Action formulae	
χ::= true	Any observable action
a	The observable action a
~ χ	Any observable action different from χ
χ & χ'	χ and χ'
State formulae	
ϕ::= true	Any behaviour is possible
~ ϕ	ϕ is impossible
ϕ & ϕ'	ϕ and ϕ'
Aγ	For each of the possible executions γ
Path formulae	
γ ::= G ϕ	Always ϕ
[ϕ{χ}U{χ'}ϕ']	At any time χ is performed and also ϕ, until χ' is performed and then ϕ'

Proper fault hypothesis processes are used to enable or disable the fault occurrences according to the previous cases.

Let us consider the formalisation of the properties in ACTL. For correct nodes, the combination of the *Agreement* and *Validity* properties (see Chapter 2, Section 2.2.1.4) can be rewritten as: "whenever the internal value of the node P is equal to 1 (action !psend_vp_1), for any execution of the processes, the nodes eventually agree on such a value (actions !vp_ofp_eqto_1, !vp_ofq_eqto_1, !vp_ofr_eqto_1). This is expressed by the following formula:

```
AG[!psend_vp_1] (A[true{true}U{!vp_ofp_eqto_1}true]
        & A[true{true}U{!vp_ofq_eqto_1}true]
        & A[true{true}U{!vp_ofr_eqto_1}true] )
```

Analogously, the three nodes must agree on value 0 as well (actions !vp_ofp_eqto_0, !vp_ofq_eqto_0, !vp_ofr_eqto_0). Similar formulae must be written for the values sent by the other nodes Q and R.

Let us consider now the case in which the node R is faulty. In this case, the ACTL formula which represents the *Agreement* and *Validity* requirement can be split in two sub-formulae: one indicating agreement for the value vp of P and vq of Q independent on the result of the vote of node R; the other expressing the agreement of P and Q on the value error related to the internal value of R.

If vp is equal to 0, we have to prove the following formula:

```
AG[!psend_vp_0] (A[true{true}U{!vp_ofp_eqto_0}true]
        & A[true{true]U{!vp_ofq_eqto_0}true])
```

Similar formulae are written for the internal value vq of Q.

The formula stating agreement on the value error for the internal value of R is:

```
AG[!rsend_vr_1| !rsend_vr_0]
        (A[true{true}U{!vr_ofp_eqto_error}true]
        & A[true{true}U{!vr_ofq_eqto_error}true])
```

Applying the model checker tool, we observe that:

1. The formulae are satisfied by the model of the system when a node can exhibit an unpredictable behaviour with the constraint that messages sent by non-faulty nodes cannot be undetectably altered by relaying nodes (assumption of authentication).
2. Some formulae are not satisfied if a node violates the assumption of authentication.

We can use the counter-example facility of the model checker AMC to find a path that falsifies the formula:

```
|= why
why: (AG [!"psend_vp_0"]
A[true{true} U {!"vp_ofp_eqto_0"}true] & .........) false 0

       0:  labelled by: !"psend_vp_0" which satisfies: !"psend_vp_0"
   20598:  (A[ true {true} U {!"vp_ofp_eqto_0"} true ] &
                     A[true{true}U{!"vp_ofq_eqto_0"} true ])(1) is false
   20598:  (A[true{true}U{!"vp_ofp_eqto_0"} true ])(1) is false
   20598:  labelled by: t which satisfies: true or tau
  ............
   25207:  labelled by:!"vp_ofp_eqto_error" which satisfies: true or tau
   31119:  labelled by: t which satisfies: true or tau
  ............
     876:  labelled by: "stop_mc" which satisfies: true or tau
END
```

This counter-example shows a path in which the action !psend_vp_0 is executed and then the action !vp_ofp_eqto_0 is not executed. Instead the action !vp_ofp_eqto_error is found, expressing the fact that P, which is non-faulty, after the voting has produced the value error for the value vp.

8.4.3 Interactive Consistency Mechanism with Four Nodes

Given the validation effort done for the 3-node case, we could expect that the validation for the 4-node case exploits the already built model, by performing some kind of incremental extension. This would avoid having to generate the global state space from scratch. Unfortunately, this is not possible due to the structure of the mechanism itself: validation has to be repeated on each instance of the interactive consistency mechanism generic definition.

The validation of the interactive consistency mechanism to cover the 4-node case therefore requires writing a new specification of the nodes of the network. At the beginning, we have written the specification applying the same methodology followed in the 3-node case. The increment of the number of nodes, the greater number of messages exchanged during any phase of the algorithm, and the new variables necessary to record the received messages and their decoded values, lead to a very large global model of the behaviour of the mechanism.

This is a typical state explosion problem case, and indeed the standard tools in the JACK environment fail to work on such a large model for lack of memory. We have followed two different strategies to solve it. On one hand, we have used a model checker that employs methods of compactly representing the state space by an implicit description based on Binary Decision Diagram (BDD). In particular, a model checker for ACTL, based on symbolic algorithms and implicit representation of the labelled transition relation of LTSs [Fantechi *et al.* 1998], has been integrated in a new version of JACK, named JACK2.

On the other hand, we have simplified the definition of the 4-node interactive consistency mechanism to a single basic protocol, where one node acts as transmitter and three nodes as receivers. This is possible since, on the basis of the work done for the 3-node case, we deduce that:

1. The global model of the mechanism is based on the interleaving of three independent basic protocols, each one following the ZA algorithm.
2. Verification is required in the project under the assumption of independent faults.
3. The occurrence of a fault in a node will alter the working of the protocol in the node itself. Such a fault will be detected by the basic protocol running on the other nodes as manifest or arbitrary fault.
4. Faulty nodes are not considered in the *Agreement* and *Validity* properties. Moreover, *Validity* is not required for the value sent by a faulty node.
5. Every basic protocol has its own set of actions, disjoint from those of the others. Formulae stating *Agreement* and *Validity* for the value sent by a transmitter node include only actions from one of the previous sets.

As a consequence, *Agreement* and *Validity* properties referred to different transmitters can be proved on the relative basic protocol.

Following the above assumptions, the specification of the 4-node interactive consistency mechanism is composed of 64 processes. Note that if we consider the complete specification of the 4-node protocols, we have 160 processes and an estimated $2*10^9$ states.

Table 8.4 shows some figures about the size of the 4-node case generated models in terms of number of states of the LTSs under some fault assumptions.

Agreement and *Validity* formulae are not verified in the case of an arbitrary faulty node which violates the assumption of authentication. This asserts that the 4-node interactive consistency mechanism is able to reach *Agreement* and *Validity* even in presence of two faults, provided that the assumption on the authentication of the messages is not violated. In the case of such a violation, even a single arbitrary fault is not tolerated.

8.5 Fault Treatment Mechanism

The fault treatment diagnosis mechanism (see Chapter 4) is based on the interactive consistency mechanism, which provides valuable data about how a node can detect a faulty remote node. In each node, the diagnosis mechanism elaborates an error syndrome which contains cumulative error statuses (based on α-counters) regarding the other nodes, by observing the interactive consistency communication exchanges. These local error syndromes are then exchanged and voted.

Table 8.4 — Number of States of some Models Generated for the 4-Node Case

Model	Number of states
Node without faults	428
Network of 4 nodes without faults	3479
Network of 4 nodes with an arbitrary faulty node which violates the assumption of authentication	122767
Network of 4 with a symmetric faulty node and an arbitrary faulty node	109613

The mechanism is decomposed in three phases: *accumulation, consolidation* and *diagnosis*. During the *accumulation phase*, each node elaborates its local error syndrome by detecting erroneous messages. During the *consolidation phase*, each node computes the global error syndrome by exchanging and voting the local error syndromes received from the other nodes, in order to reach a distributed consensus about faulty nodes. During the *diagnosis phase*, each node diagnoses the status of the other nodes using the global error syndromes.

The validation approach followed for the fault treatment mechanism separately addresses the different phases of the mechanism, trying to assess the impact they have on the overall correctness and completeness properties, even though completeness can be proved only for given classes of faults.

8.5.1 Accumulation Phase

For the *accumulation phase*, we distinguish the *detection* of erroneous messages from the *management* of α-counters. *Correctness* and *Completeness* of the detection, with respect to the class of considered faults (omission and corruption of interactive consistency mechanism messages), is verified by model-checking using the same methodology of the interactive consistency mechanism. In particular, the detection is modelled by adding, on top of the interactive consistency mechanism, processes that detect the occurrence of faults and that signal such occurrences to the environment. The detector processes follow the algorithms described in [Rabéjac 1997].

In case of faults detected in the interactive consistency protocol exchange, they emit increment actions, which are supposed to increment a counter external to the specification. Modelling increments as external signals makes the fault detection observable. In particular, we made the following changes to the model of the interactive consistency mechanism:

1. Addition of the accumulation phase processes.
2. Introduction of increment actions to denote an accusation from a node against another node.

3. Modification of the specification of the node to detect errors during the voting phase of the interactive consistency mechanism.
4. Addition of the actions to specify the beginning and the end of a cycle of the accumulation phase (the accumulation phase is done cyclically), and the termination of the interactive consistency protocol.
5. Introduction of a specific process to control one cycle of the accumulation phase (α-cycle).

The signals emitted toward the environment are considered as triggering the α-counters, which are not modelled at this stage. The α-cycle process allows the behaviour of the mechanism to be studied under different assumptions about the α-cycle (number of interactive consistency protocol exchanges that are executed before terminating the accumulation phase).

The number of states of the LTS describing the behaviour of the detection is $2*10^5$ for the 3-node case.

We prove the following properties for the accumulation phase:

- *Correctness:* if a fault has been detected, the fault actually occurred.
- *Completeness:* any occurrence of fault is detected.

In the specification, the occurrence of any increment action is visible from outside as the action inc. Similarly, the occurrence of any kind of fault is visible from outside as the action fault.

- *Correctness:*

  A[true{~!inc}U{!fault}true]
  an increment cannot occur if a fault has not occurred

  AG[!inc](~E[true{~!fault}U{!inc}true])
  between two increments there is always a fault

- *Completeness:*

  AG[!fault](AF<!inc> true)
  when a fault occurs, then there is an increment

  AG[!fault](~E[true{~!inc}U{!fault}true])
  between two faults there is always an increment

8.5.2 Consolidation Phase

The *consolidation phase* is applied on each node to achieve a globally consistent diagnosis about the node's health. This phase consolidates the local syndromes accumulated by the nodes. Starting from the results on the verification of the interactive consistency mechanism, we can prove that the *consolidation* of local syndromes through an interactive consistency network exchange does not alter correctness and completeness properties. This is a direct consequence of the fact that *Agreement* and *Validity* properties hold for the exchange of the local syndromes.

8.5.3 Diagnosis Phase

During this phase, each node can take a decision about the correctness of other nodes. *Correctness* and *Completeness* of the diagnosis algorithm over the consolidated syndromes are proved again by model-checking. The diagnosis algorithm is specified as a network of processes. The model of the diagnosis algorithm has 115 states for the 3-node case.

The following properties are proved:

- *Completeness*: all faulty nodes are identified.
- *Correctness*: any node that is diagnosed as faulty is indeed faulty.

To prove the above properties, we distinguish the case in which a node detects itself as faulty (internal detection) from the case in which a node detects other nodes as faulty (cross detection). In the following, !a_pq_F denotes the accusation by node P that Q is faulty; !p_faulty denotes the observable detection of a fault in node P.

- *Completeness of internal detection*

  AG[!a_pp_F]AF<!p_faulty> true
  whenever !app_F occurs, then P is faulty

- *Completeness of cross detection*

  AG[!a_qp_F][!a_rp_F]AF<!p_faulty> true
  whenever !a_qp_F and !a_rp_F occur, then P is faulty

  AG[!a_rp_F][!a_qp_F]AF<!p_faulty> true
  whenever !a_rp_F and !a_qp_F occur, then P is faulty

- *Correctness of internal and cross detection*

  ~E[true{~!a_pp_F &~!a_qp_F&~!a_rp_F} U {!p_faulty} true]
  there exist no computation in which the node P is detected as faulty but none of the actions !a_pp_F, !a_qp_F, !a_rp_F have occurred before

  AG[!a_qp_F]~E[true{~!a_pp_F&~!a_rp_F} U {!p_faulty} true]
  whenever !a_qp_F occurs, then P is not faulty if !a_rp_F or !a_pp_F have not occurred

  AG[!a_rp_F]~E[true{~ !a_qp_F& ~!a_pp_F} U {!p_faulty} true]
  whenever !a_rp_F occurs, then P is not faulty if !aqp_F or !app_F have not occurred

The steps above prove the *Correctness* of the fault treatment mechanism and its *Completeness* with respect to single permanent omission and corruption faults at the interactive consistency exchange level.

This definition of the covered class of faults is conservative since transient faults that occur frequently enough to be recorded and not forgotten by the α-count mechanism are covered as well. So, the definition of the class of covered faults is strongly dependent on the α-count parameters. Moreover, in the implementation of the GUARDS architecture, the α-counters are activated by other fault detection mechanisms as well, therefore extending the actual class of faults covered by completeness.

8.6 Multi-level Integrity Mechanism

The multi-level integrity mechanism implements the multiple levels of criticality model presented in Chapter 6 and defines an object-oriented integrity policy. The framework of the model is a programming paradigm for safety-critical systems in which it is possible to assign a level of confidence (*integrity level*) to each object of an application. These integrity levels distinguish components with respect to their criticality: a critical component should have a high level of confidence, while a non-critical one may have a low level. The integrity levels are used to guide the process of validating an application: low integrity level components will need a smaller validation effort.

The mechanism implementing the model consists of a run-time system guaranteeing that the co-operation among different components does not affect the overall confidence of the application, i.e., that a non-critical component does not corrupt a critical one.

Multiple-Level Objects (MLOs) are objects whose integrity level can be dynamically modified, and are a core concept of the multiple levels of criticality model. In our formalisation of this mechanisms, MLOs are described by means of CCS-Meije term, abstracting on the functionality of the objects, and concentrating only on the most relevant events — the object invocations — which trigger the change of the object integrity level [Semini 1998].

Once we have abstracted from the functional aspects, the process algebra description has a low complexity, and the generation of the model to be checked does not suffer the state explosion problem. The number of states of the models for the considered validation cases is less than 50.

The given CCS-Meije description is parametric with respect to the integrity levels of the objects. Validation has to address the interaction between objects. Model-checking is able to exhaustively verify a given property on a non-parametric finite state model. We have therefore to choose some representative cases, whose validation results can then be generalised. The considered validation cases are:

1. *Nested invocations:* A system defined as the composition of two objects, one with integrity level equal to 2, the other equal to 0. It is proved that if the functional behaviour of the first object impose that a request is served by invoking a request from the second object, then a request from an object of integrity level 1 cannot be served, since the nested server has a lower integrity level. This shows that a lower integrity level object cannot invoke a higher integrity level one by "hiding" inside an invocation to an accessible object. The proof is carried out by expressing this integrity property in ACTL and by checking it on the model generated from the algebraic parallel composition of the two objects.

2. *Concurrent invocations:* A system defined as the composition of two objects with integrity level equal to 2. The objects are concurrently invoked. We prove "object segregation" properties, i.e., that the concurrent invocations are completely independent from one another.

8.7 Discussion

We have presented our experience on formal validation of fault tolerance mechanisms conducted within the GUARDS project. This experience has shown the feasibility of model-checking techniques to formally validate the selected mechanisms.

The experience has also shown the limits of model-checking. In particular, model-checking behaves well when the structure of the system to be validated is fixed. This has occurred in the interactive consistency and fault treatment cases, because we could fix the number of nodes, which is one of the genericity parameters of the GUARDS architecture. The formal verification of generic definitions of algorithms, rather than of their particular instances, requires theorem-proving techniques and tools: this is what has been done inside the project for the clock synchronisation algorithm. In the case of the multiple level integrity policy, the system to be validated is composed of a variable, and possibly infinite, number of objects. Proofs made by model-checking aimed to validate the interactions between objects, and have to be generalised by proper reasoning. This is in our opinion one way of combining theorem proving and model-checking techniques: model-checking has the role of automating and speeding up sub-proofs in a theorem-proving approach to formal validation, as exemplified also in [Gong *et al.* 1998].

On the other hand, on non-parametric systems like our instances of interactive consistency and fault treatment mechanisms, model-checking could take advantage of using theorem-proving to validate ad hoc abstraction techniques that are often necessary to deal with state explosion problems. The time and resource constraints of the GUARDS project have not permitted us to actually address and apply the complementarity of the two approaches. We plan to exploit this complementarity in the near future to gain a larger industrial applicability of formal methods to cope with the ever more challenging validation requirements of future safety-critical computer applications.

Chapter 9

Dependability Evaluation

This chapter summarises the dependability evaluation efforts carried out within the GUARDS project. The three-pronged modelling and evaluation study encompasses the three complementary viewpoints already identified in Chapter 1, Section 1.9.2, namely: focused, abstract, and detailed modelling. They are successively described in the following sections.

9.1 Focused Modelling

One relevant part of the effort made on modelling concerned models either on selected dependability mechanisms, or on generic or specific architectural features. Here the modelling details and levels are tailored to fit the needs of the specific evaluation objectives. The focused models concerning the analysis of the α-count mechanism [Bondavalli *et al.* 1997a, Grandoni *et al.* 1998]) constitute an example on how validation analysis proves to be useful in the design process of the mechanism itself. The other models that are briefly described in this section deal with specific features for selected instances: intra-channel error detection for the railway prototype [Amendola & Marmo 1997] and phased mission systems for the space prototype [Bondavalli *et al.* 1997b, Bondavalli *et al.* 1998b].

9.1.1 α-Count

Several members of the α-count family of mechanisms (cf. Chapter 4, Section 4.1) have been modelled and evaluated. Their behaviour has been analysed and a detailed exploration of their effects on the system has been conducted. Here we report some of these models and analyses on the basic single threshold scheme [Bondavalli *et al.* 1997a] and on the double threshold variant [Grandoni *et al.* 1998, Bondavalli *et al.* 2000]. We recall that a score variable α_i is associated to each not-yet-removed component i to record information about the errors experienced by that component. α_i is initially set to 0, and accounts for the L-th judgement as follows:

D. Powell (ed.), A Generic Fault-Tolerant Architecture for Real-Time Dependable Systems, 157–191.

$$\alpha_i(L) = \begin{cases} \alpha_i(L-1)+1 \text{ if component } i \text{ is perceived faulty during execution } L \\ K \cdot \alpha_i(L-1) \text{ otherwise} \end{cases}$$

A generic component is supposed to behave according to the following assumptions:

- All fault-related signals (triggering of an error detection mechanism, completion of a testing routine, etc.) occur at discrete points in time; two successive points in time differ by a (constant) *time unit* (or *step*). For each component, a default "I'm alive and well" *state-indicator* signal is issued at each step, if not overridden by real error signals.
- The hardware directly supporting the discrimination mechanisms is fault-free, while the signal carrying the component's judgement is correct with a probability c.
- A component may be affected by permanent, intermittent or transient faults. During each time unit it may be affected at most by a single internal fault. Therefore, a component experiencing an intermittent fault may also be hit by a transient fault, but not by a permanent fault.
- Permanent, intermittent and transient faults are exponentially distributed. Therefore the probabilities of their occurrence in a time unit are constant, and will be denoted by q_p, q_i and q_t respectively. A transient fault lasts only for one step, while a component affected by an intermittent fault repeatedly gives rise to an error with a constant probability q at each step.

Table 9.1 summarises the parameters, the symbols used and their default values. The values for $q_p + q_i$ and q_t are derived assuming that the rate of intermittent or permanent fault be 10^{-4}/hour, of transient faults 10^{-3}/hour, and time discretised into 1/100 hour units.

Table 9.1 — Symbols, Definitions and Default Values

Symbol	Definition	Value
q_t	probability of transient fault per execution	10^{-5}
$q_p + q_i$	probability of permanent or intermittent fault per execution	10^{-6}
q	probability of activation of an intermittent fault per execution, given that the component is affected by an intermittent fault	0.1
c	probability that a signal on the components behaviour is correct	$1\text{-}10^{-5}$
K	the ratio by which α_i is decreased after a success	
α_T	threshold for identifying a component as affected by a permanent/intermittent fault (single threshold)	
α_H	higher threshold in the double-threshold scheme	
α_L	lower threshold in the double-threshold scheme	

The following measures are used to quantify the performance of the mechanism, with respect to the goal stated in Chapter 4, Section 4.1:

- The expected value D of the total elapsed time, after the (permanent or intermittent) fault occurrence in the component, for which it is maintained in use and relied upon, because its condition has not yet been recognised.
- *NU*, the fraction of component life in which a healthy component is not effectively used in the system. *NU* is a measure of the penalty inflicted by the discrimination mechanism on the utilisation of healthy components.

To evaluate D and *NU*, the behaviour of each individual mechanism has been modelled by means of two Stochastic Activity Networks (*SAN*) [Sanders & Meyer 1991, Sanders *et al.* 1995]. In SAN models, real-valued variables, like α_i, need to be represented by a discrete approximation which can be chosen as close to α_i as desired at the price of increasing the time and computational effort necessary to obtain the solution.

Figure 9.1 shows the plots of D and *NU* for different values of α_T and varying K for the single-threshold scheme. While increasing values of K improve D and worsen *NU*, the opposite effect is observed regarding α_T. For low values of K, D markedly improves as K increases, up to a region around a value that we will call K_D ($K_D = 0.99$ in the case of Figure 9.1); for $K > K_D$, variations become smaller and smaller. D increases for increasing values of α_T, becoming less sensitive to α_T above the region around K_D. *NU* grows slowly for low values of K up to a region around a value that we will call K_{NU} (K_{NU} is about 0.95 in the case of Figure 9.1), whereas it becomes worse and worse for $K > K_{NU}$, moreover it is very sensitive to the value of α_T. If $K_D < K_{NU}$, values of K in the interval $[K_D, K_{NU}]$ determine values for D and *NU* close to their optimum. It appears thus that, if no other constraints are given, values for K should be chosen in such interval. Unfortunately, the interval [*KD*, *KNU*] is not always identifiable, as is the case for the set of parameter values that generated Figure 9.1. In general, since a low *NU* and a low *NU* and a low D are conflicting objectives, tuning of parameters for a specific case can be done once the designer has given constraints on the desired behaviour of the mechanism, e.g., "D must be optimised while *NU* must take values lower than a given threshold". Further analyses have been performed on the sensitivity to external parameters like the fault rates and the probability c [Bondavalli *et al.* 1997a].

Figure 9.2 compares the double threshold scheme with the basic one, with K in the range [0.85, 0.999]. For the single-threshold scheme, $\alpha_T = 2$. Two curves are given for the double-threshold scheme, with different settings of α_L and α_H. The double-threshold scheme has better performance in almost all the range of K. Setting $\alpha_L = 0.8$, $\alpha_H = 3$ gives better D and lower *NU* in the range [0.94, 0.998]. For $\alpha_L = 0.4$, $\alpha_H = 2.5$, the range of improvement is [0.86, 0.997].

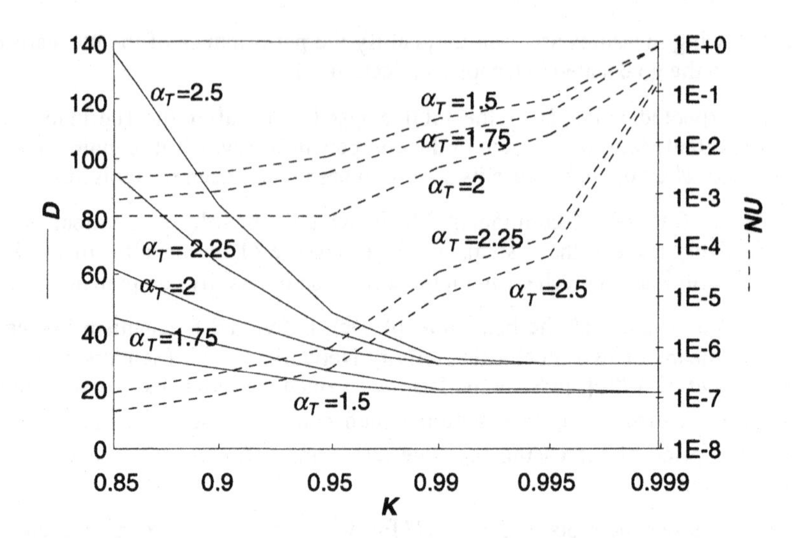

Figure 9.1 — Values of *D* and *NU* for the Single Threshold α-count

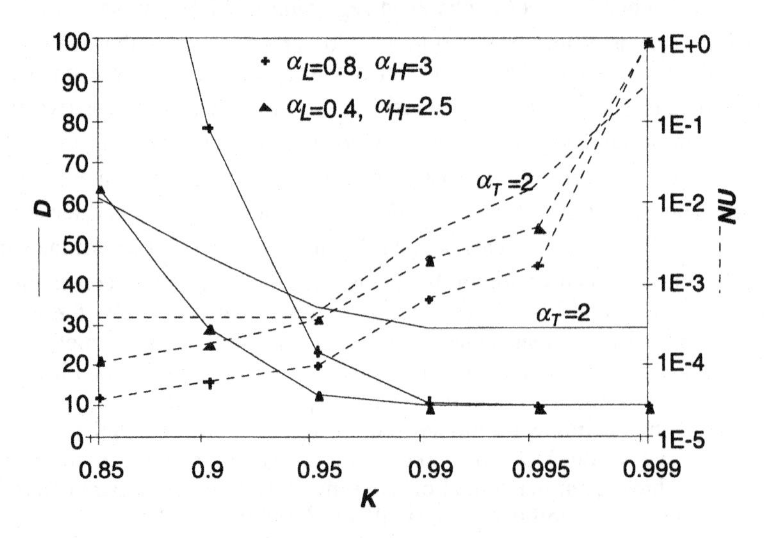

Figure 9.2 — Comparisons of Single and Double Threshold α-count

Additional analyses [Grandoni *et al.* 1998] have shown that, for all values of *K* and α_T, the double-threshold α-count outperforms the single-threshold mechanism. A proper setting for α_H and α_L has to be chosen and, obviously, the range $[\alpha_L, \alpha_H]$ always encompasses the value of α_T.

The last analyses performed have been concerned with i) the tuning of the double threshold parameters, and ii) the modelling and evaluation of the behaviour of the two schemes under less restrictive assumptions: increasing rates of intermittents (using a discrete Weibull distribution) and admitting bursts of transients. These are not reported here for the sake of space [Bondavalli *et al.* 1998a].

9.1.2 Intra-Channel Error Detection

This section describes dependability analyses carried out by Ansaldo Segnalamento Ferroviario (ASF) on their railway signalling systems [Amendola & Marmo 1997]. The objective of this study was the evaluation of the Hazardous Failure Rate (HFR) of the systems, using highly conservative hypotheses.

Among the possible causes of a failure in such systems, the most important was deemed to be latent errors that emerge simultaneously, generating matching output errors. A higher latency of such errors increases the probability of hazardous system failure. Sensitivity studies with special attention to this latency were thus conducted.

Possible failures in 2/2 and 2/3 architectures have been considered and the behaviour of a 2/2 system has been modelled. The model considers the generation of faults, their possible evolution into errors, the different error detection and recovery mechanisms built into the system, and the interactions between the two independent units used for data comparison. The evolution of failures is modelled until a final state is reached (SAFE or UNSAFE).

Faults, that hit the system at frequency L_f, are either permanent (P_{per} = 10%) or transient (P_{trn} = 90%). All transient faults are assumed to lead to errors. Some errors are detected due to channel coverage (C_{trn}), while undetected errors (1-C_{trn}) stay latent (with latency $1/L_t$) until they are processed. Permanent faults are detected by the channel software diagnostic tests with probability (C_{per}).

An important result, illustrated by Figure 9.3, is the following: when classes of errors with different rate and latency are taken into account, latency affects the system (HFR) more than the relative fault rate. In Figure 9.3, two classes of errors A and B are considered with latency (A) = 10* latency (B). By varying the relative rate of class A from 10% to 90%, it can be seen that HFR remains relatively constant, so the contribution of class B is very small. However, the different curves — which correspond to different values of latency (A) — are quite far apart, so HFR is very sensitive to latency.

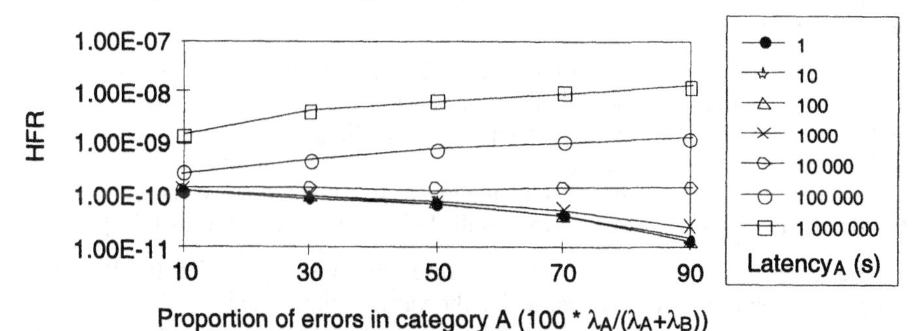

Figure 9.3 — Effect of Latency on Hazardous Failure Rate

The relationship between the system HFR, latency and fault coverage is shown in Figure 9.4. A linear dependency on the latency is observed for low values of coverage, while for high values the HFR decreases sharply.

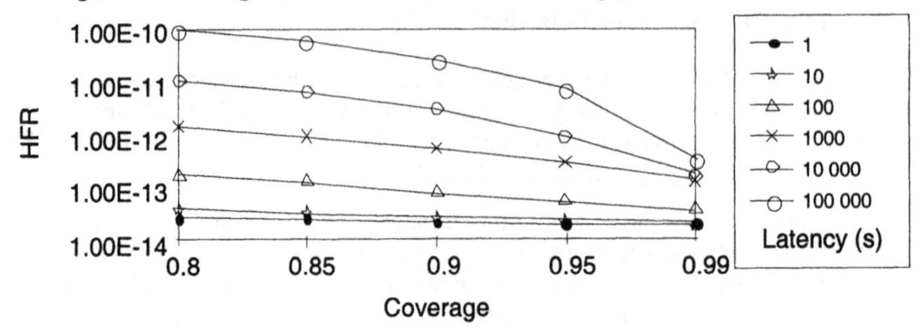

Figure 9.4 — Effect of Coverage and Latency on Hazardous Failure Rate

Other analyses reported in [Amendola & Marmo 1997] concern:

- The trade-off between the time spent for running on-line diagnostic tests and the achieved fault coverage (showing that these parameters greatly impact the system HFR).
- The effect on the HFR of the rate and distribution of faults in each unit (showing that the dependence on the fault rate is quadratic).

9.1.3 Phased Mission Systems

Many of the systems in the context of GUARDS can be classified as Phased Mission Systems (PMS), in that their operational life-time (mission) can be partitioned in a set of disjoint intervals (phases). Not only the activities that PMS have to perform during their mission can be completely different from phase to phase, but also the dependability requirements, the operational configuration and the environmental conditions can vary from one phase to another. Due to their intrinsic complexity and dynamic structure, modelling and evaluation of PMS is a challenging activity and two different methodologies have been defined. This section is thus devoted to describe these methodologies rather than in the analysis of a specific system.

The first method is based on separate models of the different phases, which are hierarchically combined in a global model of the mission. The idea underlying this method is to take advantage of the separate modelling and solution of the phase models, and at the same time to build, at a higher abstraction level, a global model of the mission. A comprehensive explanation of this methodology can be found in [Bondavalli *et al.* 1997b, Mura & Bondavalli 1997, Mura & Bondavalli 1999].

The lower level includes a set of Generalised Stochastic Petri Nets (GSPNs) that model the evolution of the system inside the various phases and which can be very different from one phase to another. This level is then completed by a set of Phase Transition Models (PTMs), which represent the changes that may occur (instantaneously) in the system at the beginning of a new phase. PTMs map the probability distribution at the end of a phase into the initial probability distribution over the states of the next phase and can be written in terms of transition matrices. The upper level is represented by a Discrete Time Markov chain, in which states represent phases, and whose structure represents the mission.

We took as an example a space-probe system with the following system phases: launch (L), planet fly-by (P), scientific observations 1 and 2 (SO1 and SO2), interspersed with phases of hibernation (H1-H4). Table 9.2 shows the phase durations, expressed in hours.

Table 9.2 — Space-Probe Mission Phase Durations

Launch	$t_L = 48$ (2 days)	Sc. Obs1	$t_{SO1} = 240$ (10 days)
Hib. 1	$t_{H1} = 17520$ (2 years)	Hib. 3	$t_{H3} = 43800$ (5 years)
Planet	$t_P = 168$ (1 week)	Hib. 4	$t_{H4} = 44040$ ($t_{H3} + t_{SO1}$)
Hib. 2	$t_{H2} = 26280$ (3 years)	Sc. Obs2	$t_{SO2} = 480$ (20 days)

Figure 9.5 describes the mission model for this example. Transition probability $p_{i,j}$ represents the probability that after phase i, the PMS either executes phase j, or fails, if $j = F$. Notice that after having completed a phase, the PMS can dynamically choose which phase has to be performed next. This choice depends on the current state of the system at the end of the phase. The transition probabilities of the model are obtained from the solution of the lower level models.

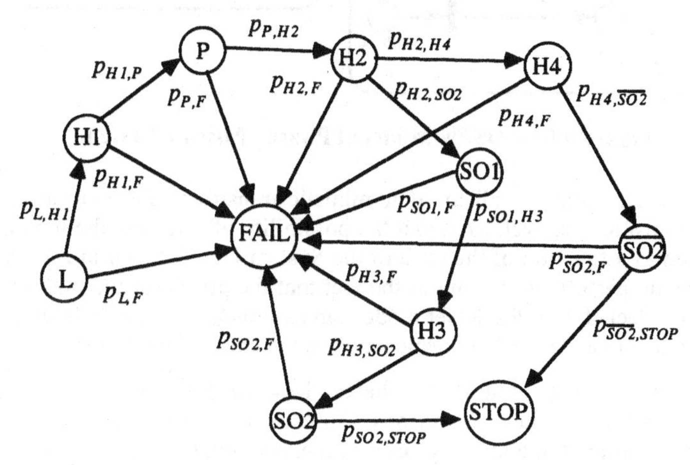

Figure 9.5 — Upper-level Model of the Phased Mission System

The solution starts from the lower level of the hierarchy. First, each of the GSPN models of the phases is translated into its underlying Markov chain. Then, according to the ordering of the phases in the mission profile, each phase model is solved to find the transient state probability distribution of its Markov chain, with an initial state probability distribution obtained from the potential preceding phase. The solution of the phase model gives the state probability distribution at the end of the phase. These final probabilities are mapped into the initial probabilities of the next phase by using the PTM. The probabilities of starting the new phase or of not surviving the current one are returned as parameters to instanciate the upper level model. Once the upper level model is fully instanciated it can be solved to obtain the probability of successfully completing the mission.

The second method [Bondavalli *et al.* 1998b] allows a single model of the overall PMS. It is based on Deterministic and Stochastic Petri Nets (DSPNs) and their underlying stochastic process, the Markov Regenerative Process. The analytical solution of the DSPN model can be partitioned and reduced to the sequential solution of each phase, requiring the same computational cost as needed by the former. A single DSPN model is defined, shown in Figure 9.6, for the same example. The model is actually composed of two logically separate models: the Phase Net, which represents the mission profile, and the System Net, which represents the system itself.

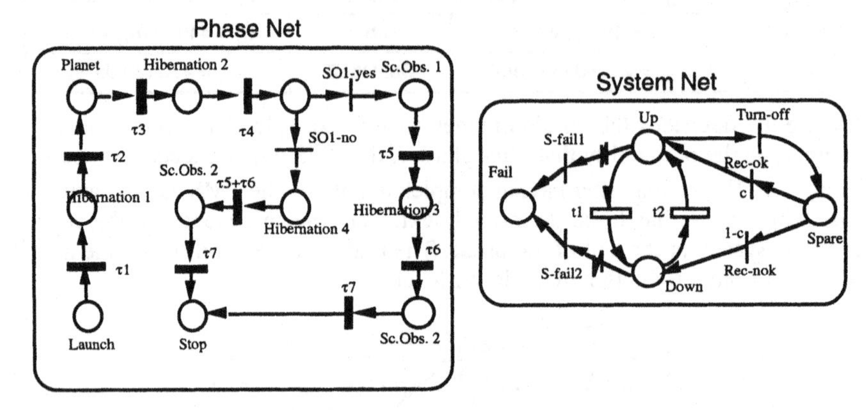

Figure 9.6 — DSPN Model of Phased Mission System

The Phase Net contains all the deterministic transitions and may also include immediate transitions as well to model the possibility of selecting the next phase to be performed as a function of the state of the System Net. A token in *Pi* means that phase *i* is being performed. Owing to the fact that the profile of the mission can be dynamically selected, the Phase Net model can assume a tree-like structure, with all the leaves linked to a place STOP, which models the end of the mission.

The System Net, built as a Stochastic Reward Net model, contains only exponential and immediate transitions. Stochastic Reward Nets have a high expressiveness, through the use of guards on the enabling of transitions, of marking-dependent transition rates and halting conditions, which can be specified by using predicates on the marking. The transition firing rates and guards for the model of Figure 9.6 are represented in Table 9.3. All these modelling features result in a very compact model. The System Net includes a FAIL place to model the failure of the phase, corresponding to a failure of the mission.

The DSPN approach improves over some weak points of the separate modelling one. In particular, it relieves the necessity of explicitly handling phase dependencies, a task that can be cumbersome and error-prone for large models. The DSPN approach automatically solves the problem of phase dependencies. A set of branching probability matrices, automatically obtained from the reachability graph of the model, map the probability distribution from one phase to the next. Due to the high expressiveness of the SRN modelling paradigm, the single DSPN model is a very concise and easy-to-read one. Moreover, since the phase-dependent behaviour of the system is described with a simple list of marking dependent predicates, it becomes very easy to rearrange and modify the model of the PMS. As already pointed out, all these features are coupled with a very efficient solution technique, which basically reduces the computational cost needed to solve the overall model to the cheaper problem of separately solving the PMS inside the various phases. Further advances could be obtained by replacing the DSPN with Markov Regenerative Stochastic Petri Nets (MRSPNs). These are Petri nets characterised by the fact that the underlying stochastic marking process is a Markov Regenerative Process, allowing missions to be modelled with generally distributed phase durations.

Table 9.3 — Transition Firing Rates and Guards for the DSPN

Transition	Hibernation	Operational	Stop
t1	λ_1#(Up)	λ_2#(Up)	0
t2	μ#(Down)	μ#(Down)	0
S-fail1	#(Up)+#(Spare) = 0	#(Up)+#(Spare) < 3	FALSE
S-fail2	#(Up)+#(Spare) = 0	#(Up)+#(Spare) < 3	FALSE
Turn-off	#(Up) ≥ 3	#(Up) ≥ 4	FALSE
Rec-ok	#(Up) ≤ 1	#(Up) ≥ 2	FALSE
Rec-nok	#(Up) ≤ 1	#(Up) ≥ 2	FALSE
SO1-yes	#(Up)+#(Spare) ≥ 4		
SO1-no	#(Up)+#(Spare) ≤ 3		

In the same example, both methods provide the same results. We analysed the system unreliability, defined as the probability that the main phase of the mission (SO2) is not successfully carried out. Figure 9.7 shows the probability of mission loss for different values of the failure rate λ and of the repair rate μ (c, the probability of successful spare insertion, has been fixed to 0.9995). For these values the final probability is very sensitive even to small changes of parameters λ and μ. As the fault occurrence rate assumes smaller values, the effect of the repair becomes less relevant. Further evaluations have been performed assuming different failure rates in the different phases and considering a range of values of the spare insertion probability c. These are not reported for lack of space.

9.2 Abstract Modelling

In addition to the focused modelling efforts described in the previous section, which were aimed at analysing specific mechanisms, a more abstract but still comprehensive modelling framework has been defined and used for the comparative assessment of the dependability of various instances of the architecture [Powell et al. 1998b, Arlat *et al.* 2000].

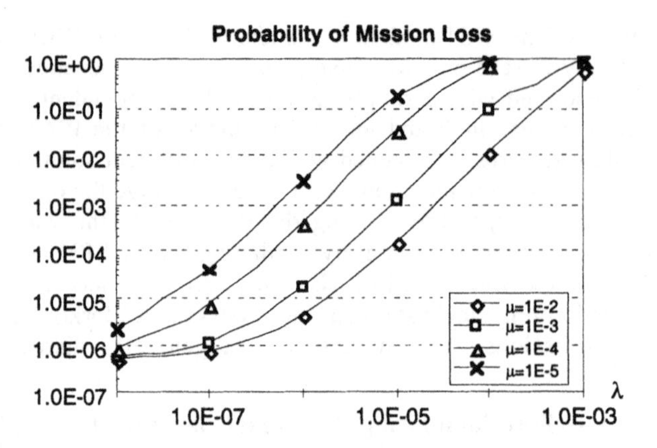

Figure 9.7 — Unreliability Analysis of the PMS

9.2.1 Modelling Framework

The modelling framework elaborates on an abstract view of the components and layers that compose the architecture. We also identify the main assumptions, the fault types and the fault-tolerance features common to the instances considered.

9.2.2 The Modelled Components

The aim is to model and evaluate the behaviour of the instances that can be derived from the GUARDS generic architecture in face of both physical faults and design faults. The granularity of the modelling is selected in terms of the possible effects of faults and the independence assumptions that can be made.

From the modelling viewpoint, the generic architecture can be described by four basic layers (Figure 9.8), which, from the bottom-up, are:

- Hardware layer: mostly off-the-shelf physical components, potentially affected both by physical faults and design faults.
- Executive layer: off-the-shelf operating systems on each processor.
- System layer: GUARDS-specific components supporting the fault-tolerance mechanisms.
- Application layer: end-user application programs — this layer is divided into sub-layers corresponding to the I integrity levels.

The layers are partitioned horizontally to account for the two dimensions of redundancy included in the architecture:

1. A set of C channels.
2. A subdivision of each channel into:
   - A set of L lanes.

- A channel nucleus corresponding to the ICN manager of that channel, the ICN link by which it sends information to the other channels, the channel I/O interfaces, etc.

Since there is no explicit means to tolerate design faults affecting the executive and system software supported by the ICN manager, the channel nucleus is considered as a monolithic component from the modelling viewpoint.

An *instance* of the architecture is characterised by the values of C, L and I, and a corresponding *fault-tolerance strategy* (two instances of the architecture could have identical values of C, L and I, but different fault-tolerance strategies). The fault-tolerance strategies that are envisaged allow instances to be degraded by passivating channels, lanes or integrity levels. Therefore, the current *configuration* of an instance may be defined by a vector of three state variables, $\{ic, im, ii\}$, with:

- $ic \in [1,C]$ indicating the current number of active channels (initially C),
- $im \in [1,L]$ indicating the current number of active lanes (initially L),
- $ii \in [1,I]$ indicating the current lowest active integrity level (initially 1).

In the three instances considered, we have C=2, 3 or 4, L=1 or 2, and I=2.

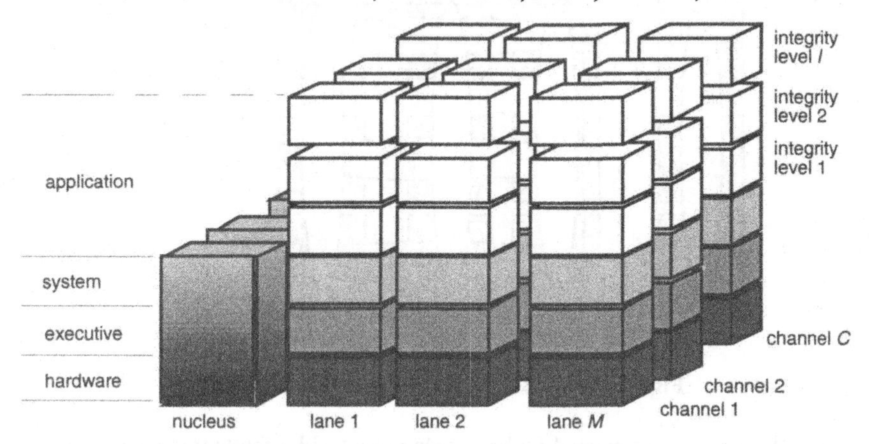

Figure 9.8 — Generic Architecture Abstract Modelling Viewpoint

9.2.2.1 Independent and Correlated Faults

Although the architecture can support diversification across channels, this was not pursued during the project (cf. Chapter 1, Section 1.2.3). We therefore assume here that the fault rates of the various components are identical across channels.

The architecture considers both physical faults (that can only affect hardware, i.e., the nucleus and the hardware layer) and design faults (that can affect both hardware and software, i.e., the nucleus and any layer). Physical faults are assumed to occur independently on different components. Some design faults can also manifest themselves as if they were independent faults, even on identical components, if the pointwise conditions necessary for their activation occur independently on the different components (i.e., Heisenbugs [Gray 1986]). We therefore consider independent faults not only at the hardware layer, but also at each software layer. We define, for each channel, ψ_N to be rate of independent faults affecting the nucleus, and $\zeta_X(m)$ to be the rate of independent faults affecting lane m, software layer X, where $X \in \{H, E, S, A_i\}$ (hardware, executive, system, integrity level i of the application).

However, common-mode failures due to correlated design faults must also be considered. Here, we consider the following possible correlated faults (Figure 9.9):

- Channel correlated faults (i.e., across all lanes).
- Lane or nuclei correlated faults (i.e., across all channels).
- Global correlated faults (i.e., across all lanes and all channels).

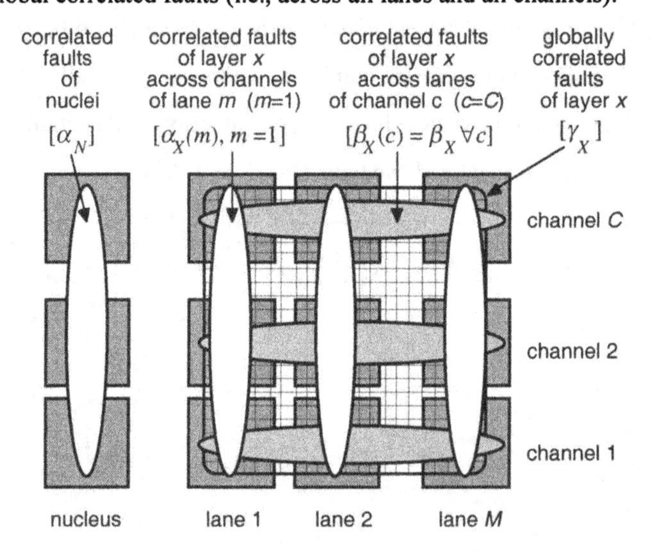

Figure 9.9 — Summary of Correlated Faults

Other than global correlated faults, we do not consider faults that simultaneously affect a processor in channel c, lane l, and another processor in channel $d \neq c$ and lane $m \neq l$ (i.e., we do not consider "diagonal" correlated faults).

The notion of correlated faults depends of course on the particular configuration of the architecture. For example, there can be no "correlated" faults within a channel if the configuration only has a single lane. So, to be able to represent different configurations and compare different instances of the architecture on a common basis, attention has been paid to the identification of rates of independent vs. correlated faults. Table 9.4 depicts the respective impact of these parameters on the independent/correlated faults for various configurations.

Table 9.4 — Modelled Events and Relationship to Basic Fault Rates

Modelled Event	i_C=1 & i_m=1	i_C=1 & i_m>1	i_C>1 & i_m=1	i_C>1 & i_m>1
Nucleus				
Independent fault on channel *c*	$\psi_N + \alpha_N$	$\psi_N + \alpha_N$	ψ_N	ψ_N
Correlated fault			α_N	α_N
Layer $X \in \{N, H, E, S, A_i\}$				
Independent fault on channel *c*, lane *m*	≡ Global correlated fault	$\zeta_X(m) + \alpha_X(m)$	$\zeta_X(m) + \beta_X$	$\zeta_X(m)$
Correlated fault of lane *m* (i.e., across all channels)			≡ Global correlated fault	$\alpha_X(m)$
Correlated fault of channel *c* (i.e., across all lanes)		≡ Global correlated fault		β_X
Global correlated fault	$\gamma_X + \alpha_X(m) + \beta_X + \zeta_X(m)$	$\gamma_X + \beta_X$	$\gamma_X + \alpha_X(m)$	γ_X

9.2.2.2 Temporary and Permanent Faults

Faults can be classified as either temporary or permanent, according to whether or not their presence is related to particular point-wise conditions [Laprie 1995]. Fault treatment depends on whether the fault is diagnosed to be a permanent fault or a temporary fault, according to the conclusion of a self-test carried out on a channel after it has been isolated from the pool. If a fault is diagnosed to be permanent, an explicit repair action must be carried out, whereas in the case of a temporary fault, certain fault-treatment strategies may authorise automatic re-integration of the faulty component.

Therefore, we need to be able to distinguish the proportions of faults that are temporary or permanent. To do so, we must consider whether the faults are independent or correlated, and whether they are of physical origin or due to erroneous design.

Independent faults of the nuclei or of the hardware layer may correspond to either physical faults or design faults. However, independent faults of the executive, system and application layers can only be design faults. Also, by definition, design faults leading to *independent* manifestations must be temporary faults, since their manifestation is related to point-wise system activity conditions. Consequently, for the nuclei and the hardware layers, we have considered the proportion of independent faults that are permanent as an important parameter of the model.

Correlated faults may be considered as either temporary or permanent, according to the fault-treatment and maintenance policy that is adopted.

9.2.2.3 Common Fault-Tolerance Features

Besides the specific fault-tolerance strategies inherent to each target instance, it can be assumed that the generic architecture supports a set of fault-tolerance features that can be considered on a common basis.

In particular, each channel has intra-channel or local mechanisms that provide error detection in addition to that provided by inter-channel error processing. The efficiency of the local error-detection mechanisms depends on the corresponding error source (Nucleus, Hardware, Executive, System, Application integrity levels) and is characterised by coverage values c_X, $X \in \{N, H, E, S, A1, A2\}$.

Inter-channel error processing depends on the number of operational channels. As long as there are at least three operational channels, any errors due to a single faulty channel are assumed to be masked (and detected) by majority voting. In the case when two operational channels are available, a two-out-of-two vote is considered and single channel errors are assumed to be detected. Let the coverage of these assumptions be defined as c_{Vc}, where index $c \in \{2,3,4\}$. Since a four-channel configuration is capable of tolerating arbitrary faults, c_{V4} can reasonably be assumed to be 100%. The other coverage values should be considered non-perfect; for example, the non-coverage includes the case of a Byzantine fault leading to failure of the clock synchronisation mechanism.

Detected errors trigger fault diagnosis to determine which channel is at fault. Any faulty channel is isolated from the operational channels to execute a self-test aimed at determining whether the fault is permanent or temporary. Let the execution rate of the self-test be defined as δ and let the coverage of the self-test with respect to permanent faults be defined as c_{ST}. If the fault is judged to be permanent, it is repaired with rate μ. If the fault is judged to be temporary, the channel is reintegrated with rate ρ. With probability $1-c_{ST}$, the self-test fails (e.g., a permanent fault is erroneously judged to be a temporary fault); in this case, we (pessimistically) assume that a catastrophic failure occurs. The permanent faults represent a proportion π of the independent faults affecting the nuclei and the hardware layer.

Any correlated faults, except those occurring at integrity level 1, are assumed to lead to a catastrophic failure[1]. Correlated faults at integrity level 1 should be confined to that level by the integrity policy enforcement mechanisms. Let the coverage of these mechanisms be c_{IP}. Detected violations of the integrity policy cause the incriminated layer to be discarded.

All undetected errors are assumed to lead to catastrophic failure.

9.2.3 The Target Instances

Three instances of the GUARDS architecture were modelled and evaluated, one for each end-user application domain: railway, nuclear propulsion and space.

---

[1] Only two integrity levels are being considered for the three target instances.

9.2.3.1 Railway Instance

The evaluated railway instance is described in Section 1.10.1 and illustrated in Figure 1.4 (Chapter 1). The nominal configuration features three channels, each including one single lane. Upon occurrence/activation of a fault, if the error is detected locally, a diagnostic-restoration procedure is immediately launched; otherwise, the error can still be handled by the two-out-of-three vote mechanism (coverage c_{V3}). If only two channels are available and if the error is detected, the system is switched to a safe state (benign failure). It is assumed that all undetected errors lead to a catastrophic failure. Also, whatever the number of channels available, the activation of a correlated fault leads to a catastrophic failure.

9.2.3.2 Nuclear Propulsion Instance

The evaluated nuclear propulsion instance is described in Section 1.10.2 and illustrated in Figure 1.5 (Chapter 1). This is a two-channel instance with two lanes per channel. Within each lane, all application components are duplicated and compared: the pair of lanes in each lane thus forms a self-checking pair able to detect errors due to faults activated independently in each lane. Consequently, the coverage values c_X, $X \in \{H, E, S, A1, A2\}$ will be higher than in the railway instance.

Apart from what was considered for the railway instance, if the error is detected locally, then the nominal two-channel architecture is degraded to a single-channel configuration. If the error is only detected by the two-out-of-two vote (coverage c_{v2}), the system is then switched to a safe state. When in the single channel configuration, the system is put into the safe state upon detection of an error. The correlated fault activation rates depend upon the number of operational channels. In the two-channel case, all correlated fault types are accounted for. In the single-channel configuration, only global correlated faults are considered. Correlated faults across channels can be detected locally, but on both channels simultaneously which causes the instance to be switched to the safe state.

9.2.3.3 The Space Instance

The evaluated space instance is described in Section 1.10.3 and illustrated in Figure 1.6 (Chapter 1). This is a four-channel instance with two lanes per channel. One lane supports the nominal application whereas the second lane runs a safety-monitoring application or a back-up application. Any detected malfunction of the nominal application causes the instance to switch to the back-up application (*survival* mode). When in the two-channel configuration, the instance operates in a two-out-of-two mode. Results of computations that are declared as error-free by the intra-channel mechanisms are compared and, in case of disagreement, the instance is put into a safe state[2]. However, if errors are detected locally, by intra-channel

---

[2] There is usually no passive "safe" state for a space application. It was nevertheless considered so as to be able to carry a comprehensive study for this instance as well. Note that the passive "safe" state is different to the "survival" mode, which still requires processing capacity so that the backup application can be executed.

mechanisms, the channel declares itself to be faulty and the instance switches to the single-channel configuration.

Due to the increased combinations of failed channels and lanes, the *space instance* features more complex degraded configurations {*ic*, *im*, *ii*}, ranging from {4,2,1} (nominal configuration) to {1,1,2} (survival mode only), leading to a more complex model. In particular, the rate of manifestation of independent faults depends on whether the configuration is bi-dimensional or has degraded to a single-channel. This applies also to faults affecting the nucleus.

9.2.4 Modelling Strategy

The modelling strategy is inspired from the modular modelling approach described in [Kanoun *et al.* 1999, Kanoun & Borrel 2000]. First, a global model is built made of abstract subsystem models or *sub-models*; these sub-models are linked by dependency models. Second, each subsystem model is refined as a Generalised Stochastic Petri Net (GSPN).

9.2.4.1 Sub-models

Although they are derived from the same generic architecture, the three instances considered differ by:

- The number of hardware and software components involved that ensure both the required services of the system in nominal (fault-free) operation and the fault-tolerance functions.
- The global strategy for fault tolerance (and maintenance).
- The considered correlated faults (see Figure 9.9), which in turn influence the fault-tolerance strategy.

Three main types of sub-models have thus been distinguished (Figure 9.10). The first sub-model is that of a channel; the global model comprises one such sub-model for each channel. The second sub-model represents the fault-tolerance strategy. The last one models the effects of correlated faults.

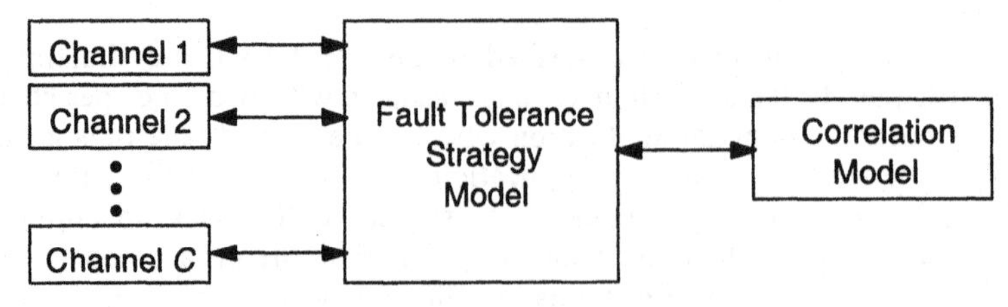

Figure 9.10 — Main Sub-models

9.2.4.2 Example of a Sub-model

Due to page-length limitations, it is not possible to present here the sub-models for all considered instances. Nevertheless, as an illustration of the type of model used,

Figure 9.11 depicts the channel model corresponding to the railway instance. Channel models designed for the other two instances are similar.

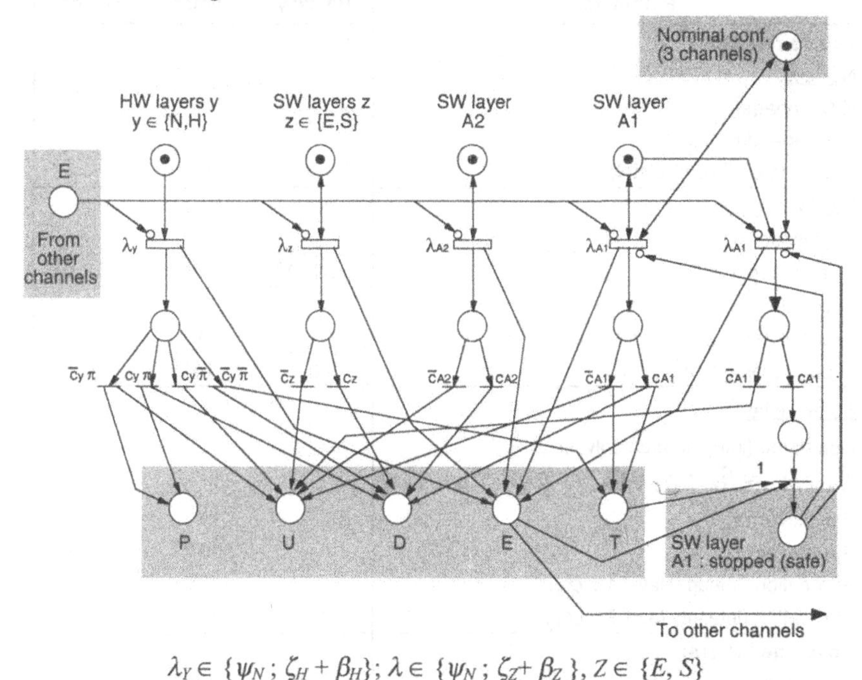

$$\lambda_Y \in \{\psi_N ; \zeta_H + \beta_H\}; \lambda \in \{\psi_N ; \zeta_Z + \beta_Z\}, Z \in \{E, S\}$$

Figure 9.11 — Sub-model for each Channel (Railway Instance)

The model distinguishes between physical and design faults. The distinction between permanent and temporary faults applies only to hardware (nucleus and processors); all design faults are considered to be temporary. Also, permanent faults are assumed to be activated only once (i.e., the consequence of activating one or several permanent faults is the same). Conversely, we assume that many temporary faults can be activated.

An important feature of each sub-model is constituted by the "interfaces" linking it to other sub-models. Four such interfaces are shown as shaded areas on Figure 9.11 (see Table 9.5 for an explanation of the notation). In particular, one interface exhibited by the channel model gathers data concerning channel state (place E is marked whenever a fault is activated as an error), fault nature — permanent (P) or temporary (T), and if the error is detected locally (D) or not (i.e., undetected (U)). These interfaces facilitate the descriptions of the interactions of the channel model with the other sub-models. For example, the consequence of the activation of independent faults in integrity level 1 application depends on the number of available channels. When i) the nominal 3-channel configuration is no longer available (i.e., the system has degraded to a 2-channel configuration), and ii) if the error provoked by the fault is locally detected, then only layer 1 of the application software switches to the safe state. Another type of interaction between channels is as follows: the marking of place E in one channel prevents the activation of independent faults in the other channels.

Table 9.5 — References Values for the Model Parameters

Parameters	Railway	Nuclear	Space
Failure rates (h^{-1})			
Nucleus:			
ICN manager			
- independent: ψ_N		0.999 10^{-5}	
- correlated (interchannel): α_N		10^{-8}	
Hardware layer			
Processor			
- independent: ζ_H	0.99 10^{-5}	0.989 10^{-5}	0.989 10^{-5}
- correlated (intrachannel): β_H		0	
- correlated (interchannel): α_H		10^{-7}	
- correlated (global): γ_H	0	10^{-8}	10^{-8}
Software layers:			
Correlated (interchannel) only, i.e.,			
$\zeta_Z = \alpha_Z = \beta_Z = \gamma_Z = 0, \ \forall Z \in \{E, S, A1, A2\}$			
Executive: α_E		10^{-4}	
System: α_S		10^{-6}	
Application (integrity level 1): α_{A1}		10^{-5}	
Application (integrity level 2): α_{A2}		10^{-6}	
Coverage factors:			
ICN manager		0.8	
Hardware & software layers			
$c_H = c_E = c_S = c_{A1} = c_{A2}$	0.8	0.99	0.8
Interchannel voting			
$c_{v1} = c_{v2} = c_{v3}$		1	
Integrity policy: c_{IP}		0.8	
Self-test: c_{ST}		0.99	
Others:			
Proportion of temporary faults: $1 - \pi$		0.9	
Channel self-test rate: δ (h^{-1})		100	
Channel restoration rate: ρ (h^{-1})		100	
Channel repair rate: μ (h^{-1})		0.1	

9.2.5 Examples of Evaluations

In this section, we briefly illustrate the types of insights on the considered instances that can be derived from this modelling effort. The models were solved using the SURF-2 dependability evaluation package [Béounes *et al.* 1993].

We first compare the three instances with respect to the reliability and safety measures for a set of nominal values. Then we present two examples of sensitivity analysis obtained for the railway instance to assess the impact of: i) the proportion of temporary *vs.* permanent faults, and ii) the correlated faults in the software layers.

9.2.5.1 Comparison of the Instances

The considered parameter values for all three instances are shown in Table 9.5. For the software layers, independent faults are neglected and only correlated faults between channels are considered. Unless otherwise stated, these values will serve as reference for subsequent analyses. Figure 9.12 compares the reliability and safety for the three considered instances.

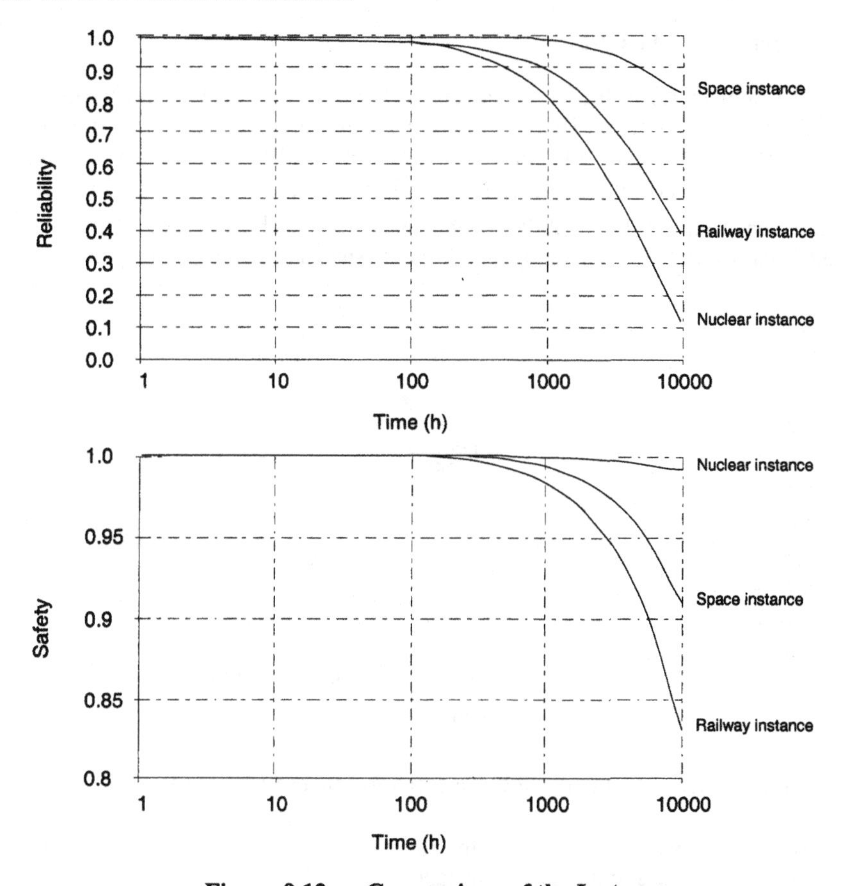

Figure 9.12 — Comparison of the Instances

The ranking of the reliability curves simply reflects the redundancy at the channel level (C = 4, 3 and 2 for the space, railway and nuclear propulsion instances, respectively). However, this ranking is modified when safety is considered: indeed, the nuclear instance is ranked first, followed by the space and railway instances. Such a ranking of the nuclear instance can be explained by the systematic intra-channel comparison between the two lanes featuring diversified executive layers, which results in:

- the choice of higher values for the coverage factors (0.99 instead of 0.8, see Table 9.5),

- the ability to detect errors due correlated faults affecting one lane across all channels.

The ranking of the other two instances reflects the fact that the railway instance does not provide any mechanisms for detecting errors due to correlated faults affecting all channels, while the space instance allows for a proportion of these to be detected, either locally, or thanks to the diversification of its executive layers.

9.2.5.2 Sensitivity Analyses

Figure 9.13 depicts the influence of correlated faults in the software layers in the case of the railway instance. The curves compare the reference case with the case where no correlated faults are considered for any of the software layers (i.e., $\zeta_Z = \alpha_Z = \beta_Z = \gamma_Z = 0,\ \forall Z \in \{E, S, A1, A2\}$). The figure clearly reveals the very significant sensitivity of the reliability of this instance to such correlated faults — as opposed to hardware faults (encompassing both physical and design faults).

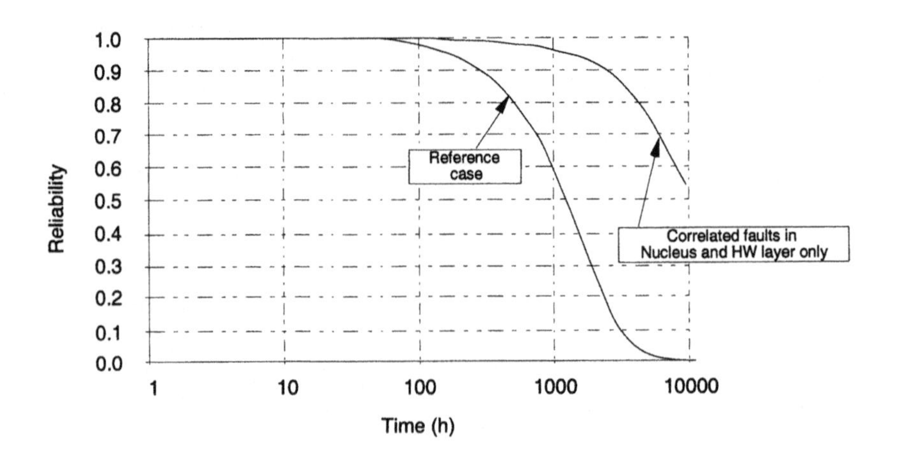

Figure 9.13 — Influence of Correlated Faults in Software Layers

Figure 9.14 analyses the impact of the ratio of permanent *vs.* temporary faults (parameter π) in the case of the railway instance.

To focus on the analysis of the physical faults, the parameter values of the reference case have been modified so that no correlated faults are considered for any software layer (i.e., $\zeta_Z = \alpha_Z = \beta_Z = \gamma_Z = 0,\ \forall Z \in \{E, S, A1, A2\}$), as for Figure 9.13. The significant variations observed are mainly due to the large difference between the (temporary fault) restoration rate ρ and the permanent fault (repair rate μ) — see Table 9.5.

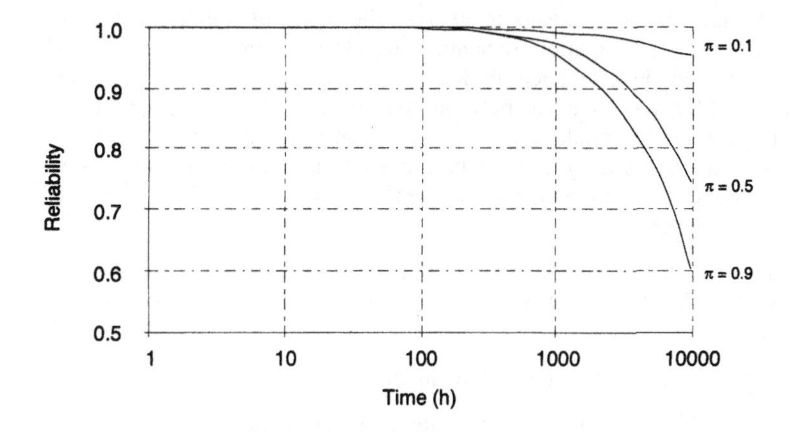

Figure 9.14 — Influence of Ratio of Permanent Faults (π)

9.3 Detailed Modelling

Detailed modelling is complementary to focused modelling (cf. Section 9.1) since it considers the whole system, and not only a specific part of it. It is also complementary to abstract modelling (cf. Section 9.2) since it reflects the system's *internal* structure in the dependability model.

9.3.1 Objectives and Strategy

Model-based evaluation methods are intended to support the system designer when configuring the generic architecture and to assist in the early and continuous assessment of the dependability of a specific instance. More generally, they aim to demonstrate the compliance of the system being built with respect to the client's dependability requirements. Hence, when considering the evaluation activity as an integrated part of the development process, modelling becomes an iterative process, performed recurrently as the knowledge concerning the system's internal structure increases. Accordingly, the *first* objective targeted by the detailed modelling activity has been to reduce the recurrent work by providing *reusable* component models allowing the evaluation of multiple design alternatives, thus reflecting the modularity of an architecture based itself on (mostly) predefined software and hardware components, and fault tolerance schemes.

The evaluation task is not restricted to the *quantification* of the various dependability attributes. It also covers the *qualitative* characterisation of a system by allowing analysis of the failure modes of the system's components and identification of their effects on the service delivered by the system (e.g., Failure Mode Effect and Criticality Analysis (FMECA)). Since qualitative analysis of the behaviour of a system in the presence of fault is usually based on a pure intellectual analysis, it shows its limits when dealing with complex hardware/software combinations. In that case, it may benefit from the means used for the quantitative analysis, both as a means to perform the analysis (e.g., *what happens if...?* analyses on a detailed model) and as a means to express the result of the analysis (e.g., characterisation of error propagation, dependencies, etc.). Accordingly, the *second* objective of the detailed modelling activity has been to provide means (methods and components) to tighten the coupling between system design and evaluation and more generally, to consider modelling as an engineering activity, carried out by system engineers not necessarily experts in dependability evaluation. This latter consideration was one of the main reasons that led us to choose stochastic Petri nets as the modelling language.

Towards the two ambitious objectives stated in the previous paragraphs, the strategy followed within GUARDS was to perform parallel investigations in two main directions:

- How to enforce the identification of the main modelling hypotheses.
- How to provide reusable, composable and refinable basic component models.

These two directions of investigations are detailed in the following sections. A third direction, devoted to the definition of a modelling environment supporting modular and hierarchical modelling was also investigated [Jenn 1998b]. For conciseness, it is not further considered hereafter.

9.3.2 Identification of the Modelling Hypotheses

A model — and the confidence one may have on the results provided by this model — is only meaningful with respect to a set of well-defined hypotheses. These hypotheses may over-simplify reality, and may be incomplete or even false, but the essential point lies in the fact that they must be clearly stated.

Towards this goal, we have introduced the concept of a *meta-model* aimed at supporting the expression of these hypotheses or, stated differently, specifying the validity domain of the evaluation model. In a complementary manner, the meta-model should also provide a means to identify the hypotheses that may have a strong impact on the system's dependability.

In practice, the meta-model is based on a simple notation supporting the expression of the characteristics of an instance of the architecture, and the mechanisms, components and processes that may have an impact on its dependability. In its preliminary form, this notation is essentially focused on the expression of the *relations* that may exist between the system's components, such as:

- Relations involved in the propagation of errors (e.g., flow of data, flow of control).
- Relations induced by the fault prevention and elimination processes (i.e., all maintenance operations including low-level resets, board exchanges, software upgrades, etc.).
- Relations involved in the correlations between errors (e.g., existence of common mode fault sources) — more generally, relations related to the coverage domain of error detection/masking schemes.

- Relations induced by the activation process (e.g., execution of a software component).

Each of these relations is associated to a given operator. Expressing the modelling hypotheses becomes then a matter of specifying graphically the pertinent relations between these components. For example, consider a system composed of two channels, *C1* and *C2*, executing the same service in a 2-out-of 2 organisation (when either channel is diagnosed as faulty, the system is brought to a safe state), and configured as follows:

- The service is implemented as two interacting functions *F1* and *F2*.
- Each function *Fi* is implemented by a recovery block structure.
- Each recovery block structure is composed of one primary (*P*), one alternate, or secondary (*S*) and one acceptance test (*AT*).
- *F1* uses some information produced by F2 and vice versa.
- *F1* and *F2* are executed on the same platform.

This system is represented by the graph shown in Figure 9.15.

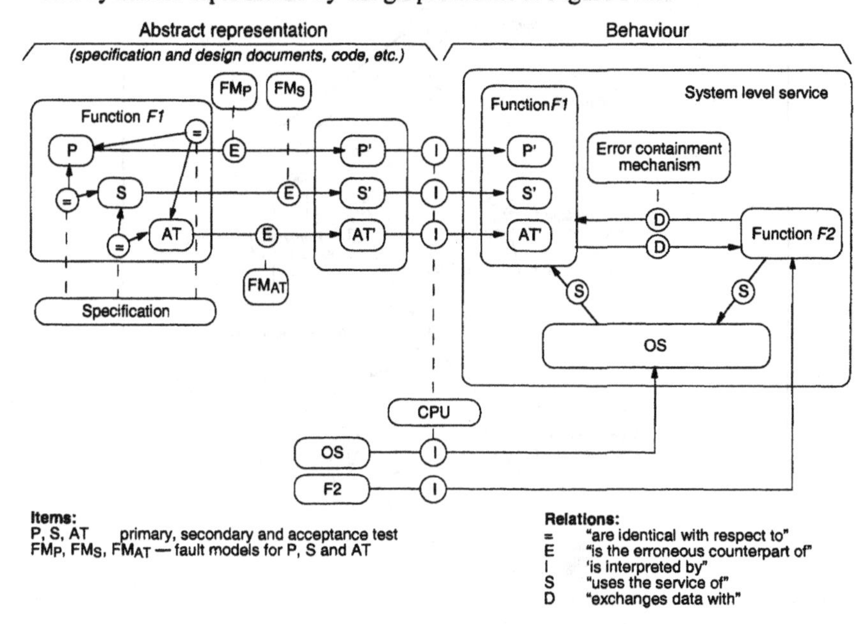

Figure 9.15 — Example of a Meta-model

This model can then be transformed — manually — into a classical dependability model, such as a Markov model. The addition or the removal of a relation is reflected in the Markov model. This is illustrated on Figure 9.16 and Figure 9.17. Another interest of such models is that of allowing the propagation of relations between components to be studied.

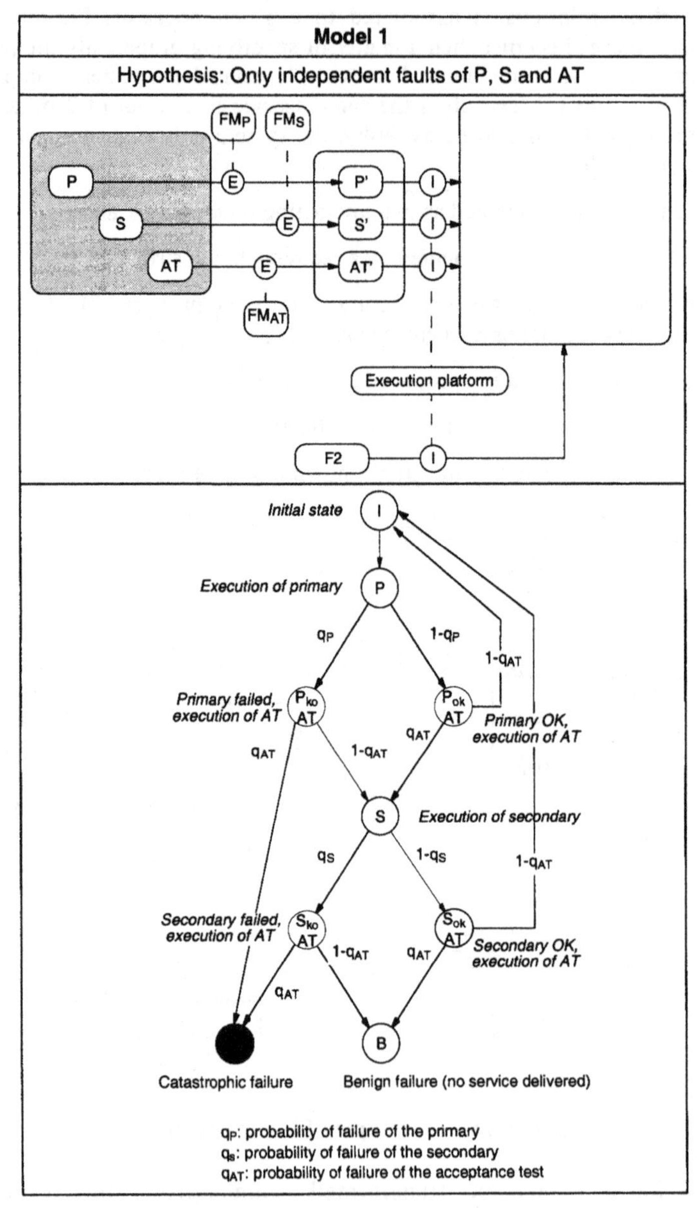

Figure 9.16 — Mapping Meta-model to Evaluation Model (*Model 1*)

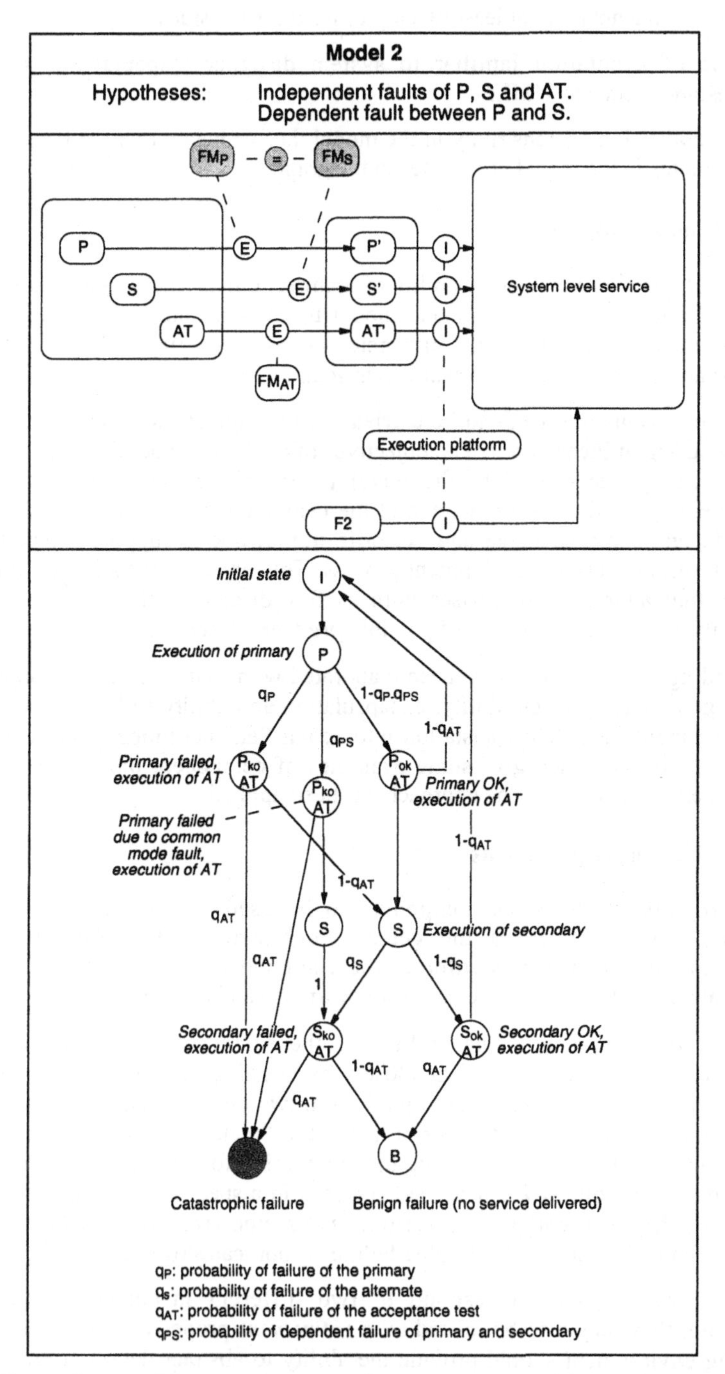

Figure 9.17 — Mapping Meta-model to Evaluation Model (*Model 2*)

The investigations carried out within GUARDS have only scratched the surface of this difficult subject. Besides the necessary improvement of the completeness and consistency of the notation, at least two topics need further study:

- Use of a notation familiar to system designer (especially a software designer), such as UML.
- Formalisation of transitivity in the model, i.e., of how a relation between two elements in the model propagates in the graph.

9.3.3 Component Models

The ability to compare the dependability of architectural alternatives depends on the ability to limit the changes from one model to another, and to impose common modelling hypotheses and, ideally, parameter values. *Genericity* and *configurability* of models are two preconditions towards this goal.

Morover, evaluation needs to be carried out throughout the development cycle and has a strong influence on it. The objectives of analytical modelling and the level of detail change according to the development phase. With respect to this development cycle, the main objective of the evaluation by analytical modelling is to provide one comprehensive and versatile framework to support dependability evaluation throughout the development process. This objective imposes a modelling approach that addresses all phases with as few discontinuities as possible, by allowing the *reuse* and *refinement* of possibly *predefined* models.

Accordingly, the models have been elaborated with four main design objectives in mind: genericity, configurability, encapsulation and ability to be refined. These four requirements and their translation into actual Petri net models are detailed in the next section. We then give some examples of component models and briefly present an example to which the approach has been applied.

9.3.3.1 Design Requirements

Genericity allows the same component to be used on various instances (in particular, two instances of the generic architecture considered as design alternatives). In practice, this enforces the component models to be focused on orthogonal and abstract aspects (e.g., error occurrence and error propagation).

Configurability concerns the ability for the same component model to be instanciated so as to correspond as closely as possible to a given actual entity. Instanciation parameters have been organised in two classes: numerical parameters (e.g., parameters of probability distributions) and "meta-models" making the link between abstraction levels. Concerning the latter point, for example, the selection of a 1-out-of-2 or a 2-out-of-2 scheme is expressed in a specific model that interprets the state of the basic components (erroneous *vs.* non-erroneous) in terms of the system dependability state (catastrophic failure *vs.* non-catastrophic failure).

If two models may correspond to two different instances of the same architecture, they may also be related by a *refinement* property. Consequently, the modelling environment should provide the ability to abstract details from detailed models and, conversely, to refine abstract models (e.g., when new information concerning the actual components of the instance becomes available).

This process is illustrated on Figure 9.18. On this figure, the model of entity B makes use of five main component models, including one devoted to the detection of errors. In a first phase, the way this error detection process is actually implemented is not considered. In the next phase, this process is considered to be performed by a specific software entity (entity B) which is itself associated with five processes, and so on.

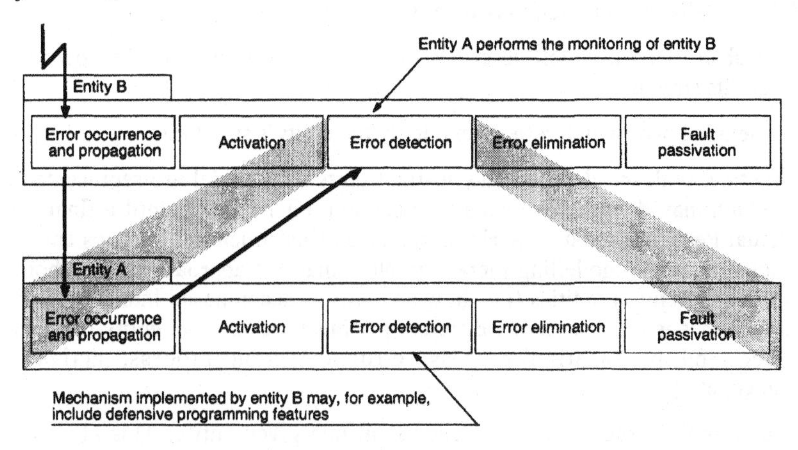

Figure 9.18 — Recursion in the Modelling

The obvious drawback of introducing multiple elementary and orthogonal component models is *complexity*. With respect to this issue, our baseline has been to transfer the complexity of building an overall dependability model — done many times — towards that of building the component models themselves — done just once. To facilitate their use, the component models have been made as modular as possible, by emphasising encapsulation and interoperability. Encapsulation has been encouraged by using "sub-nets" (i.e., Petri nets embedded in Petri nets) wherever possible. These sub-nets are used to isolate the parts of the model related to the same aspect (e.g., error activation and propagation, error detection, etc.). Interoperability has been enforced by the use of a common communication means between component models based on place aliases (i.e., references to Petri net places defined elsewhere).

The actual models have been realised with MOCA-PN [Dutuit *et al.* 1997], a Petri net modelling tool originally developed by Elf. With respect to the objectives targeted by the detailed modelling activity, this tool offers at least two interesting features:

- It supports the interactive execution of a model, which allows a qualitative assessment of the model (e.g., to see whether the transitions are fired according to the expected sequence).
- It provides a means to group sets of places/transitions and arcs to make them appear as a single unit. This is the basic mechanism used to implement the

concept of component models: a component model is a group that may be copied where required.

The tool also allows the creation of place aliases, i.e., references to places located in another group. This is the basic mechanism by which interactions between component models are implemented.

9.3.3.2 Model Design

A model is usually decomposed in two parts:

- An abstract part, which describes the behaviour of an entity independently from its structure,
- A detailed part, which reflects explicitly the entity's structure.

In practice, both the detailed and abstract parts of a model are represented by groups, which may themselves recursively contain other groups, until a final stage where actual Petri nets are necessarily introduced. This structure of groups and sub-groups constitutes the modelling hierarchy. Note that this approach is used both to express the hierarchy of entities (i.e., how an entity is decomposed into sub-entities) but also as a means to separate the different pertinent views of an entity (e.g., fault occurrence and propagation process, error detection process, activation, maintenance, etc.).

Detailed and abstract views may coexist in one given entity. This allows, for example, part of an entity to be modelled with structural details while another part is described in more abstract terms. In a redundant system, this may consist in modelling one redundant channel with a high level of detail (e.g., the channel in which a fault is supposed to be activated) while keeping the model of the other channel(s) as abstract as possible.

The coexistence of multiple abstraction levels imposes the introduction of specific components in charge of (i) elaborating the global state of the entity from the states of the detailed and abstract parts, and (ii) maintaining the consistency between the state of the various sub-components.

For example, if a component error detection mechanism triggers at the detailed level, this information is transmitted to the global model where it is combined (via an "or") with analogous information coming from other detailed and abstract component models to elaborate a global "error detection" event. This event may trigger a fault passivation mechanism, modelled at an abstract level, which may impact the state of the detailed components.

Detailed and abstract views may also correspond to two successive modelling phases of a single component. For example, during a first modelling phase, a channel may be simply modelled by an abstract "ErrorState" model limited to the error occurrence process (which *implicitly* takes into account the failures of the channel's sub-entities). Then, in a second phase, this model may be refined by taking into account the structure of the channel and the designer's knowledge concerning the failure modes of the channel's sub-entities. Note that during this refinement, care should be taken to adapt the parameters of the abstract model in a consistent manner (in order for the same item not to be taken into account twice, i.e., in both the abstract and detailed models). In practice, this could mean that the failure rates used in the abstract model need to be adapted by, for example, setting them to zero to nullify the fault process already modelled at the detailed level.

The graphs in Figure 9.19 give typical examples of what could be models for an entity providing some error detection and processing features, at two abstraction levels. On these figures, ellipses represent Petri nets and arrows represent arcs between the nets. These models illustrate one particular case where the detailled model does not provide any error detection feature, the latter being delegated to the abstract model (which then makes no distinction between sub-entities, the detection characteristics being given as a single coverage factor).

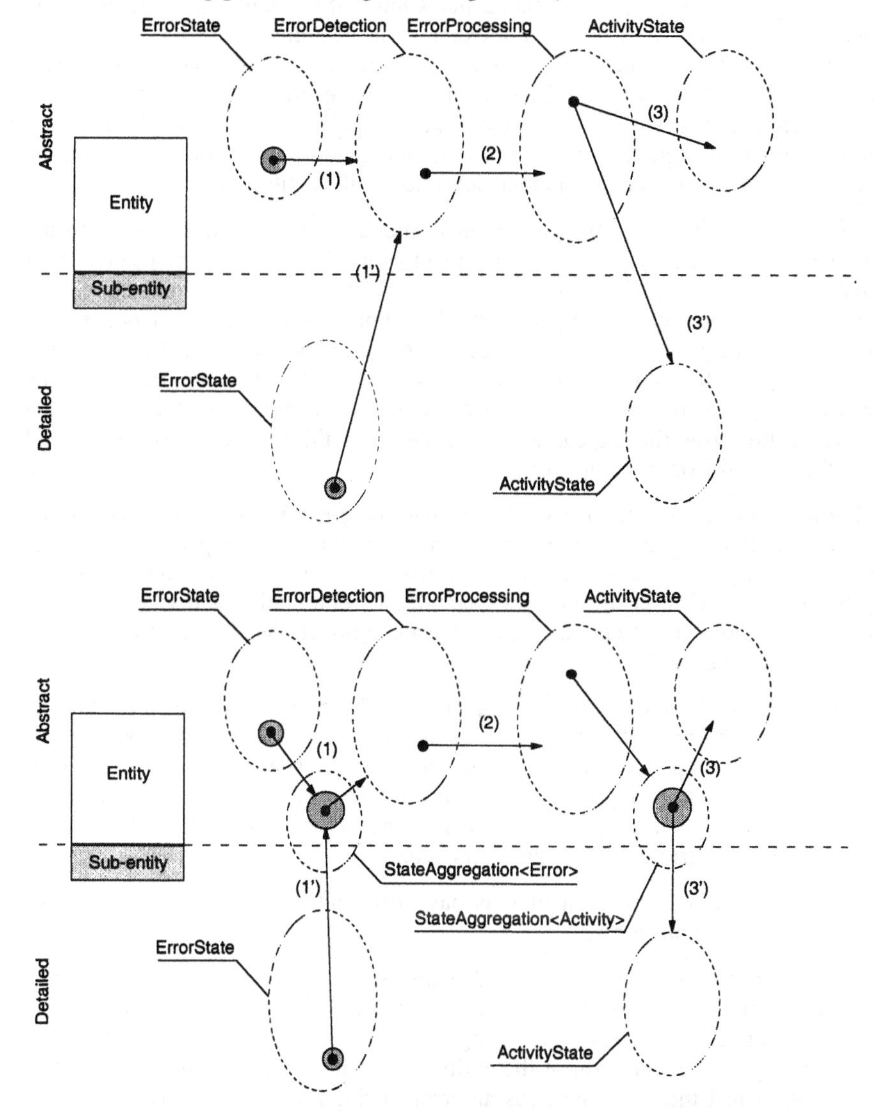

Figure 9.19 — State Decomposition and Aggregation

In the first graph, the "ErrorDetection" component checks directly the abstract and detailed "ErrorState" component (arcs (1) and (1')). If there is an error in either of them, the "ErrorProcessing" graph is "informed" and, finally, the "ActivityState" goes into the idle state until some action is performed (by the operator, for example). On this example, the (1') edge represents a bottom-up flow and the (3') edge represents a top-down flow.

In the second graph, two specific components have been introduced to maintain the consistency between the state of the component models (detailed and abstract) and the state of the entity. As previously, the information flows in both direction.

Ensuring consistency among related states implicitly means that there are some dependencies between these states. For example, if the entity is a host board and the sub-entity the CPU, or if the *Active* to *Idle* state transition actually means the power-off of the host. The solution illustrated by the second graph allows a clear identification of the entity's state (one has simply to check one place in the "StateAggregation" graph) and a simple maintenance of the consistency of the states of all components involved in the definition of the entity's state.

The decomposition into basic component models is organised according to the aspects that have an obvious impact on dependability, such as fault occurrence, fault activation, error propagation, error detection, etc. This first level of decomposition constitutes the core set of component models. For example, the *fault occurrence process* and the *fault activation process* are distinguished since the activation process is tightly linked to the component activity (whether it is idle, running some self-test, etc.), whereas the fault *may* occur independently from the component activity. In this case, the segregation is induced by a third-party component model (here, the "ActivityState" component).

Furthermore, a distinction is made between the error detection process and the error elimination process since (i) these processes may not be performed by the same entity (e.g., different independent software entities may be involved), and (ii) the detection of an error does not guarantee its elimination and vice-versa (for example, a periodic reset operation may put a component back in a non-erroneous state although no error has been detected).

The core set of components is complemented by component models regarding aspects that may impact the behaviour of the core components. For example, the "ActivityState" component (introduced to represent the state of an entity, running, idle, self-testing, etc.) may be linked with the "ErrorPropagation" component to take into account — in particular — the fact that a running application is more prone to propagate errors than an idle application.

In all, ten main component models have been introduced. Among them, the following ones are noteworthy:

- The "ErrorState" component models the fault occurrence, the error activation and the local error propagation processes. An error is considered to propagate from "ring" to "ring", a "ring" being used to model the propagation of the error at the entity's interface (i.e., an error in ring 1 may only affect the entity whereas an error in ring 2 may propagate outside the entity).

- The "DependabilityState" component links the system state and the presence of an error to the consequence on the service delivered by the system.
- The "ActivityState" component is used to represent the different states of an entity with respect to the service it delivers.

Figure 9.20 illustrates an "ErrorState" component model. Without going into detail, we will simply note that the fault activation and error propagation processes are mainly described by the right-hand part of the Petri net. This net is connected to the rest of the model via several interface points, implemented by means of place aliases (represented by double-circled places on one side and the number of the place on the other side).

Figure 9.21 presents another component, dedicated to the modelling of the error detection process performed by an entity on itself (self-checking mechanism for a hardware component or defensive programming measures for a software component). This component is "activated" when an error has propagated itself to "ring0"; this is achieved by means of a place alias (in the upper left side of the graph). It is also connected to a component in charge of taking into account common mode faults (upper right part of the graph) and to a component modelling the error diagnosis process (lower right part of the graph).

9.3.3.3 Application

Component models have been used for a preliminary model of a 2-out-of-1 instance reconfigurable to a 1-out-of-1. The modelled system features:

- Two strictly identical channels.
- Two hosts boards executing the same application software in each channel,
- A CPU, an OS and three application software component *C1*, *C2* and *C3*, with *C1* and *C2* at a different integrity level in each host.

In the 2-out-of-2 architecture, a faulty channel may be put into a safe state due to the triggering of its own self-checking mechanisms (i.e., decision taken at the intra-channel level). The instance as a whole may be put in the safe state should an error be detected at the inter-channel level (no decision can be made on the origin of the error). The intra-channel error detection is based on several mechanisms, including:

- Monitoring of the outputs (local readback of the outputs, in order to detect failures of the output or read-back boards and error in one of the host boards, the output and the readback being performed by separate host boards).
- Intra-channel comparisons (comparisons of the processing results provided by each host board).
- Intra-host error detection mechanism.
- Defensive programming techniques (control flow monitoring).
- Checking of the compliance of the data transfer with respect to the multi-level integrity policy.
- Low-level (processor and MMU) error detection mechanisms.

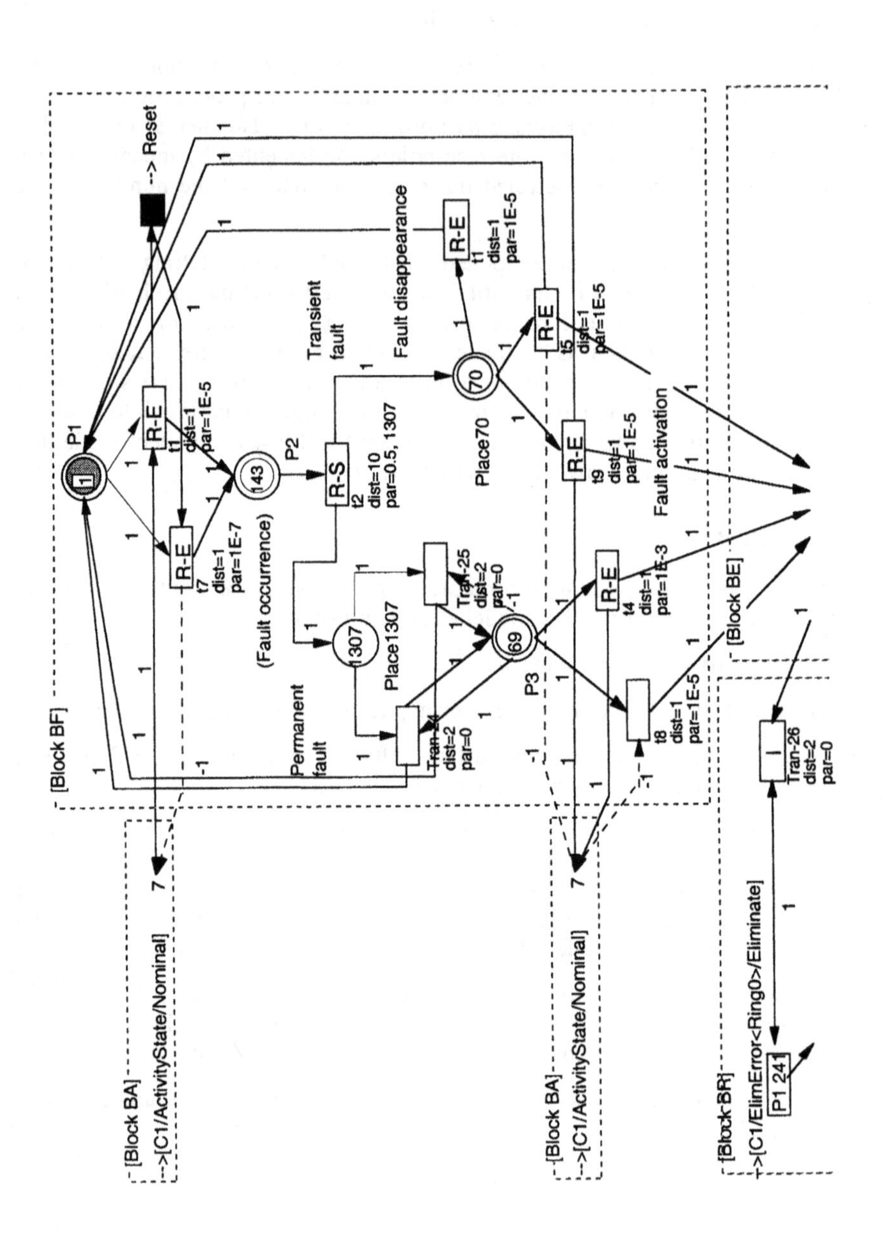
[Block BF]
[Block BA]
-->[C1/ActivityState/Nominal]
[Block BE]
[Block BR]
-->[C1/ElimError<Ring0>/Eliminate]
--> Reset
P1
P2
P3
Place70
Place1307
Transient
fault
Fault disappearance
Fault activation
(Fault occurrence)
Permanent
fault
R-E
R-S
t1
dist=1
par=1E-5
t7
dist=1
par=1E-7
t2
dist=10
par=0.5, 1307
t5
t9
t4
par=1E-3
t8
Tran-24
Tran-25
Tran-26
dist=2
par=0
P1 241

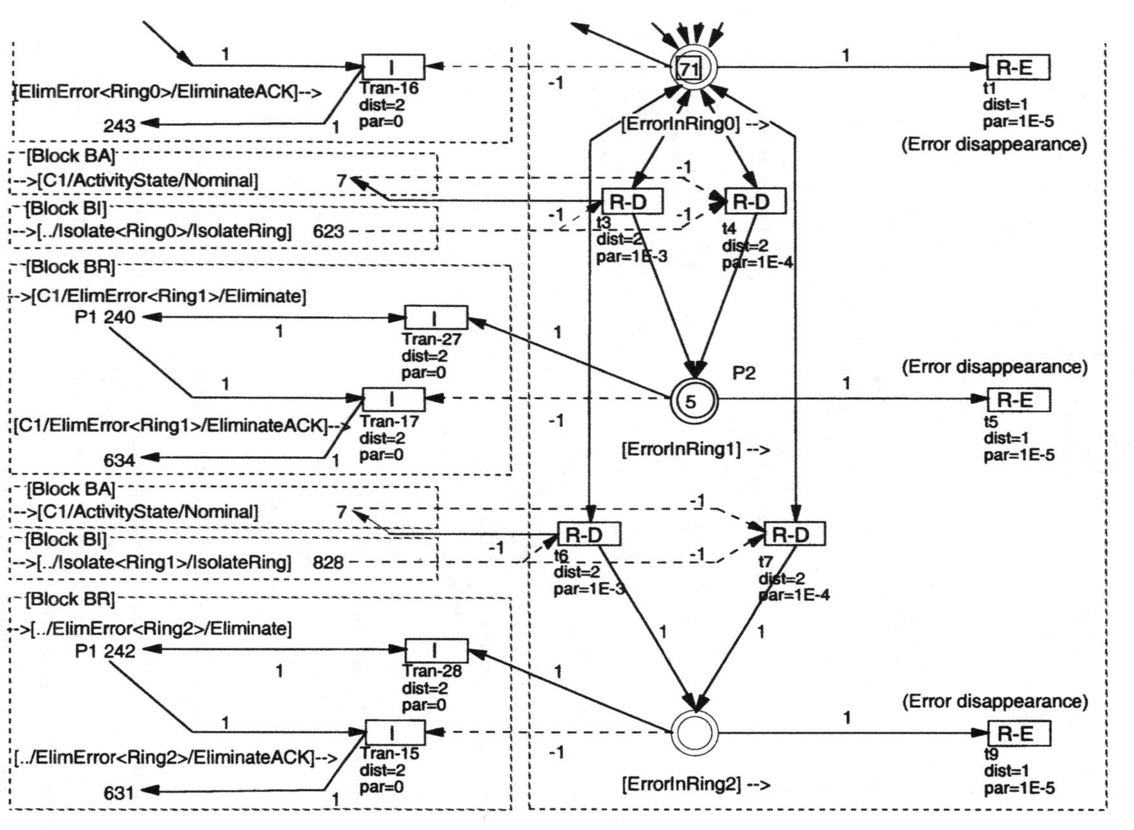

Figure 9.20 — The "ErrorState" Component Model

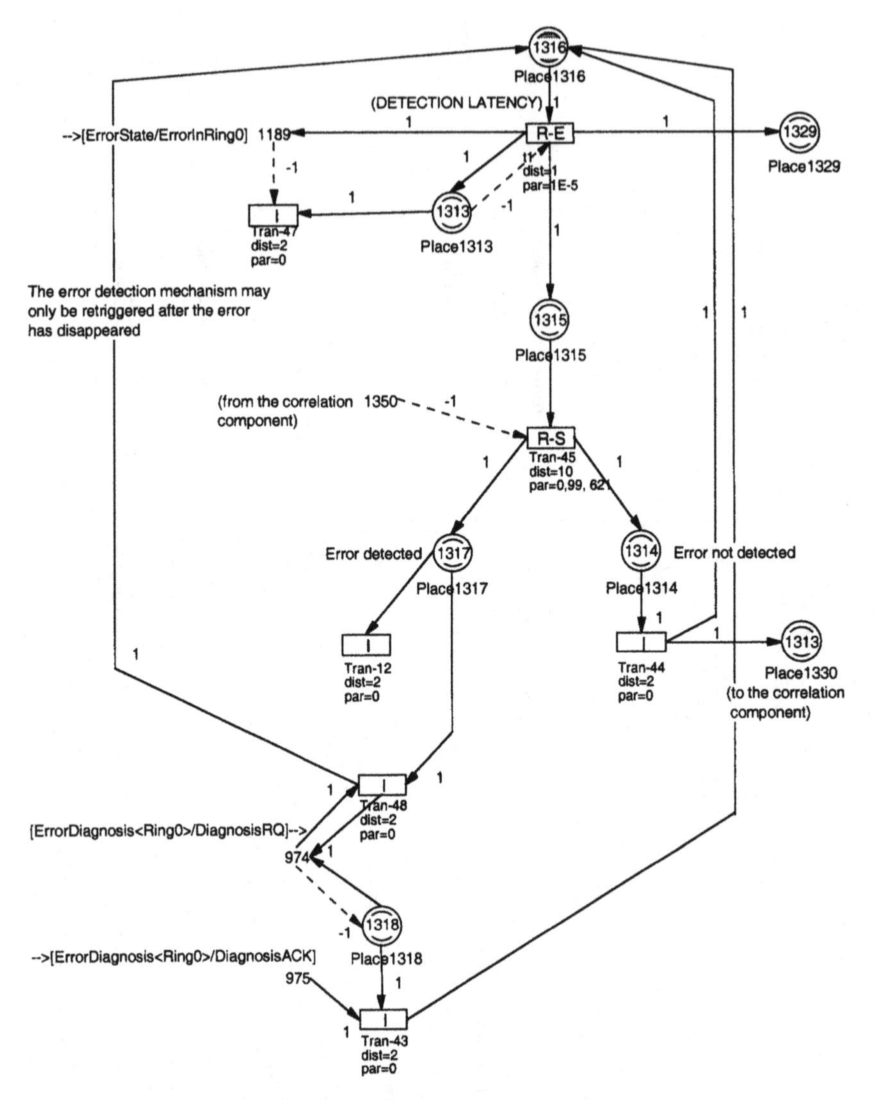

Figure 9.21 — The "LocalErrorDetection" Component

The complete model corresponding to this system is quite complex. It involves several modelling layers and mixes two abstraction levels: one channel being modelled in a very abstract way and the other taking into account details concerning its hardware and software structure.

The final model contains around 700 transitions and 500 places, organised in 5 levels of hierarchy. Only a few of these elements were introduced into the model individually, the vast majority correspond to instantiations of component models ("ErrorState", "LocalErrorDetection", etc.). Currently, however, only interactive simulations have been performed to carry out initial validation of the component models and their interactions.

The objective of providing an environment where building a model would be essentially a matter of connecting component models together has not yet been fully achieved. Currently, component models cannot be interconnected without having to partially modify them (to add edges and references to places). These modifications require a good knowledge of the internal structure of the component models, which is exactly what we were trying to avoid. This disadvantage could be removed by providing a more sophisticated means for defining modelling "modules" (i.e., sub-nets with well-defined interfaces) and some means for expressing the rules for interconnecting these modules.

9.4 Discussion

This chapter has summarised the three modelling viewpoints supporting dependability evaluation that are part of the GUARDS validation environment (see Chapter 1, Section 1.9). These three viewpoints are complementary and span focused, abstract and detailed modelling. Focused modelling addresses the analysis of selected generic dependability mechanisms (such as fault diagnosis and intra-channel error detection) and specific application scenarios (such as phased missions). Abstract modelling is aimed at the comparative assessment of complete instances of the architecture, and in particular, according to the main architectural dimensions: channels, lanes, integrity levels). Detailed modelling offers a generic support for a refined analysis of selected fault tolerance algorithms (e.g., recovery blocks). Such modelling activities proved useful in supporting and justifying the decisions made during the design of the generic architecture and its dependability mechanisms.

Furthermore, this comprehensive modelling framework constitutes a powerful design aid for developers to help them tailor instances of the GUARDS architecture to fit the dependability needs imposed by their specific application domain. Indeed, the sensitivity analyses that can be carried out using the modelling framework (for all modelling viewpoints) constitute a powerful aid to help the system developers make evaluation-based design choices. Towards this end, the abstract modelling should be applied first in order to identify the main architectural choices on the basis of the dependability measures of interest. Then, the detailed modelling framework allows for a finer dependability analysis during the design phase by accommodating the internal structure of the instance being developed. If needed, focused modelling can also be used to carry out a thorough study of specific mechanisms that were found crucial in the previous analyses.

Chapter 10
Demonstrators

This chapter first outlines the implementation of the inter-channel network manager, which is common to all instances of the architecture. Then, the demonstrators implemented for each end-user application are briefly described.

10.1 Inter-Channel Network Manager

A GUARDS instance is composed of two to four channels, each including a number of computer boards (host processors) and passive devices (peripherals) connected by a VME bus. On each channel, access to the Inter-channel Communication Network (ICN) is under the responsibility of an *ICN Manager* (Figure 10.1).

10.1.1 Architecture

The ICN manager is a standard Motorola VME162 board with a 68040 CPU. On the ICN processor board, the VME bus is connected to a dual port memory (DPRAM_1), which is also addressed by the 68040. This memory area is the main physical interface between the host software and the ICN software. It is used by the host software to:

- Drive the ICN software, by means of specific commands and flags.
- Provide private values for consolidation or broadcasting.
- Access the results of the consolidation.
- Send and receive asynchronous messages.
- Monitor the ICN software behaviour.

In addition, the ICN software can raise interrupts to the host software through the VME bus.

D. Powell (ed.), A Generic Fault-Tolerant Architecture for Real-Time Dependable Systems, 193–227.

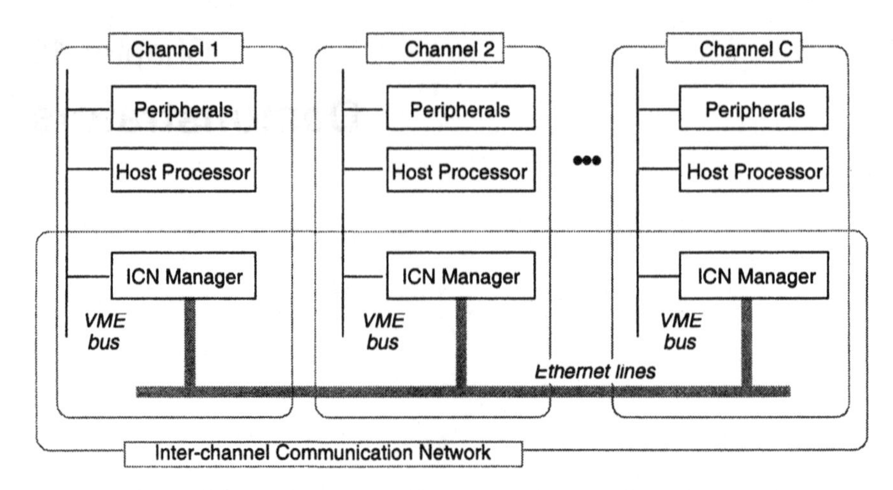

Figure 10.1 — The Inter-channel Communication Network

Each ICN manager board (Figure 10.2) is composed of the Motorola 68040 master board (ICN_M) and two Motorola 68EN360 piggy-back boards as ICN controllers (ICN_Cs). In the maximum GUARDS configuration (four channels), one ICN controller is configured as one transmitter and one receiver (ICN_C T/R), while the other one is configured as two receivers (ICN_C R/R). The ICN_Cs share the same clock frequency generator and the related pre-scaler, but they have independent timers connected to the shared pre-scaler.

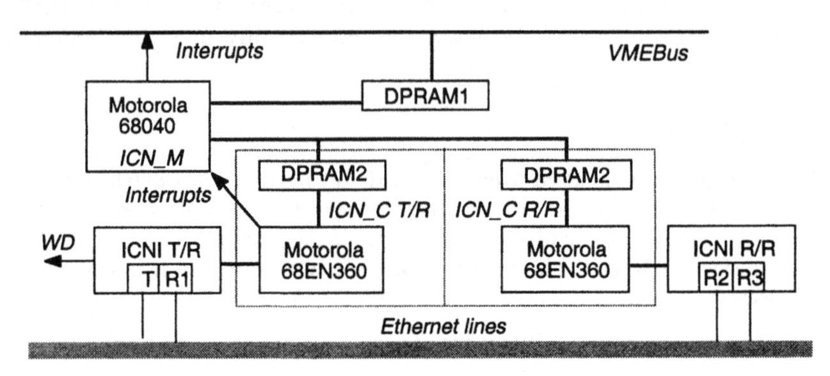

Figure 10.2 — ICN Manager Architecture (Maximum Configuration)

The interactions between the host and the ICN manager take place through the VME bus and are implemented as:

- Interrupts to the hosts generated by the ICN manager.
- Data at fixed locations in the shared memory transferred under the control of the host. The host is the VME master whilst the ICN_M is the VME arbiter. The shared memory is the DPRAM_1 directly accessible by the hosts through the VME.

The slot time period is generated by the timers of the ICN controllers; ICN_C T/R is in charge of triggering the ICN manager by a specific interrupt at each slot time. The duration of the last slot in a cycle is calculated by the ICN manager and provided to the ICN controllers. The nominal or last-slot duration is used to properly load the timers at each cycle.

An internal dual port memory (DPRAM_2) allows exchanges between the ICN manager and the two ICN controllers. DPRAM_2 is not accessible through the VME. Therefore, data has to be transferred from DPRAM_1 to DPRAM_2 (and vice versa) by the ICN manager.

The ICN controllers determine the time of reception of synchronisation messages and provide such values in DPRAM_2. The arrival time is given by the slot number and the contents of the slot period generation timer when the synchronisation message is acquired.

The interactions between the ICN manager and the ICN controller(s) take place through the DPRAM_2 data and control connections and are implemented as:

- Interrupts generated by ICN_C T/R towards ICN_M.
- Data at fixed locations in DPRAM_2.

10.1.2 Software Mechanisms

The ICN software implements a number of the GUARDS dependability mechanisms, in particular the *clock synchroniser*, the *ICN manager*, and the *inter-channel error manager*.

10.1.2.1 Clock Synchroniser

The clock synchroniser component maintains synchronisation between the instances of the ICN software running on the different channels and, as a consequence, the instances of the host software. The synchronisation is required by the data consolidation algorithms (implemented by the ICN manager component). It is assumed that related private values are exchanged at the same time between the channels over the Ethernet lines.

This is achieved by a timer on the ICN Processors that generates interrupts at a fixed frequency. The time between two interrupts is called the "slot" time. The transmission (and reception) of any piece of related information is always executed during a slot. Data items transmitted by each channel during the same slot are related to each other: e.g., private values of a data item to be voted. The time duration of a slot can be configured during ICN software initialisation.

To synchronise the slot interrupt generation on the ICN Processors, the "Resynchronisation Cycle" (abbreviated to "cycle" in the following) is defined as a number of sequential slots. The number of slots in a cycle is also a configuration parameter of the ICN software. At the beginning of each cycle, the instances of the ICN software exchange synchronisation messages. A synchronisation algorithm allows skew correction by tuning the duration of the last slot of the cycle.

The clock synchroniser component raises an interrupt to the host processors through the VME bus at the beginning of each cycle to allow the synchronisation of the host software.

10.1.2.2 ICN Manager

A typical GUARDS application task has a behaviour that can be sketched out as follows:

- Read a sensor.
- Consolidate the input, by interactive consistency over the ICN.
- Calculate the results that are to command the actuators.
- Consolidate the results, by voting over the ICN.
- Write the consolidated result.

The activity can be thought of as a transaction composed of three cyclic threads (Figure 10.3). Each thread has its own deadline and offset; they all share the same period.

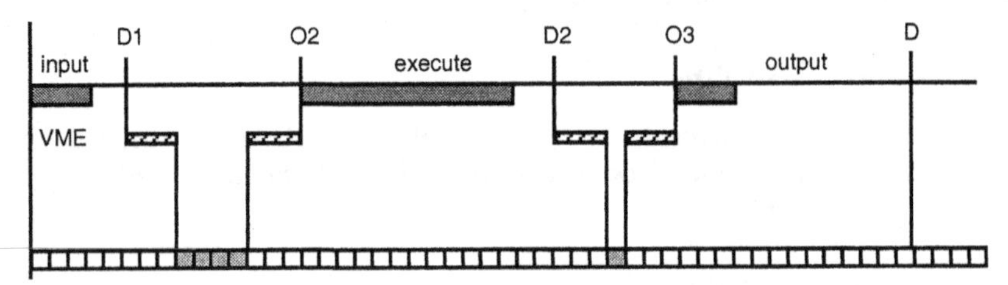

Figure 10.3 — An Application Transaction

For each input consolidation, the value is written into the DPRAM_1 memory through the VME as soon as it is available. The transmission delay over the VME is not deterministic, but we assume that it can be bounded. To be sure that data is available on each channel, the first ICN slot that can be used is after the deadline (D1 on Figure 10.3) for reading plus the VME delay. Four slots (i.e., the maximum number of channels) are used for the interactive consistency protocol.

After termination of the interactive consistency protocol, the vector of values is available to the application software in DPRAM_1. This means that the consumer cannot run before a time equal to the deadline of the producer plus twice a VME delay plus the interactive consistency duration. If we assign to the consumer an offset equal to this time (O2 on Figure 10.3), it will be able to fetch significant data. A similar explanation holds for the exchange and consolidation of the results, except that only one ICN slot is necessary since result consolidation only requires a simple majority vote rather than an interactive consistency exchange.

When a number of tasks run, they share the ICN and conflicts on the slot assignment can happen if intermediate deadlines are close to each other, as sketched out in Figure 10.4.

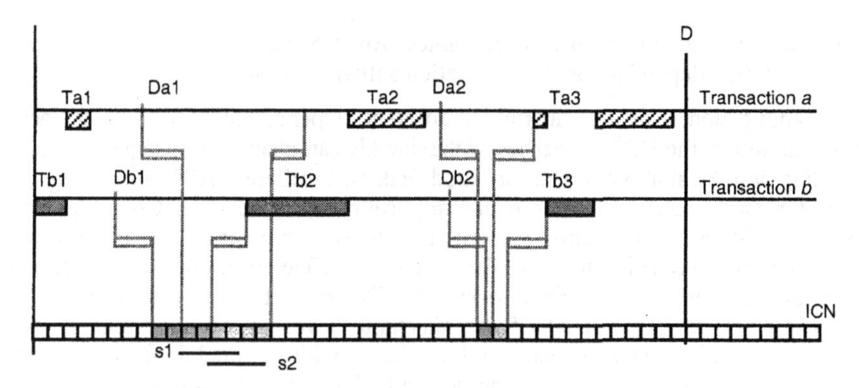

Figure 10.4 — Illustration of Slot Assignment Conflicts

For simplicity, we assume that the deadline is the same for transactions *a* and *b*. Since *Db1* < *Da1,* transaction *b* can reserve the optimal set of slots. Transaction *a* desires the slot set *s1* but some slots are already busy, so the set *s2* is assigned. The resulting offset to be applied to *Ta2* is greater than in the case of no interference. The opposite situation happens during the assignment of slots for the second consolidation.

Here we suppose that *Ta2* has a priority lower than *Tb2*, so no pre-emption takes place. The start of *Ta1* is delayed by the interferenceof Tb1.

It should to be noted that the timing properties of the threads (offsets in particular) cannot be made independent from the slot assignment. In particular, the intermediate deadlines (*Dx1* and *Dx2*) must be set and tuned according to both the schedulability analysis performed for each host and the assignment of ICN slots.

The ICN manager component is table-driven. For each slot in a cycle, a "Slot Descriptor" fully describes the operations to be performed through a number of "Directives" (up to four) for the ICN manager. Each directive causes the ICN software to:

- Move data between the DPRAM_1 and the Ethernet lines.
- Calculate and verify the checksum of the data in a slot.
- Compare related data in order to support the data consolidation mechanisms.

An appropriate sequence of slots and the related directives cause the ICN software to perform the voting and the Interactive Consistency algorithms. The operations to be performed by the ICN manager for each slot of a cycle are defined in the "Cycle Table", which contains the related slot descriptors. In the simplest case, the period of the application software transactions is equal to the resynchronisation cycle, so the slot assignment can be repeated in all cycles. In general, the transactions may have a period that is a multiple (or sub-multiple) of the ICN cycle. In this case, a slot assignment that can be repeated forever must cover a number of sequential cycles with a total duration that is a multiple of all periods. The sequence of cycles after which all ICN manager activities are repeated is called an "ICN Frame". The ICN frame is defined by the "ICN Table", which contains the corresponding set of cycle tables. An ICN table may contain one or more cycle tables depending on the application software transaction periods.

The application software can run in different "Operational Modes". For each operational mode, the HRT properties of the threads can change. Consequently, the scheduling and the slot assignment are modified, so a different ICN table must be defined for each operational mode of the application software. The ICN tables are generated off-line taking into account the timing properties of application transactions and composing threads. The aim is to define an optimal allocation of slots which maximises throughput over the ICN and software performance. To support this activity, the GUARDS Architectural Development Environment (cf. Chapter 7) provides a slot allocator tool combined with a schedulability analyser which allow the designer to check the schedulability of each software thread, to define the intermediate offsets of the transactions and to produce the related ICN table.

10.1.2.3 Inter-Channel Error Manager

The inter-channel error manager is in charge of detecting and notifying to the application software (in the internal data in DPRAM_1) errors of different types that constitute error syndromes for the local and the remote channels. Such errors are mainly related to communication over the ICN: loss of messages, corrupted messages, disagreements of the relayed messages during the interactive consistency protocol, etc.

The detected errors are used by the application software component "Pool Manager" to feed the α-count mechanism that supports the inter-channel fault diagnosis. The mechanism is based on the accumulation of errors for a given time interval. At the end of each interval, the α-count is increased according to the number and the type of accumulated errors. When there are no errors, the α-count is decreased.

Separate α-counts are maintained by the pool manager for the local channel and for all remote channels. When the α-count becomes higher than a given threshold, the corresponding channel is considered faulty (note that this is only the *local* opinion about the correctness of the corresponding channel). The diagnosis algorithm is performed by consolidating the local opinions of all channels by interactive consistency. After the consolidation, if a majority of the channels assert that a channel is faulty, then fault treatment (switch-off of the channel) is initiated.

10.2 Space Demonstrator

10.2.1 Logical Fault Tolerant Architecture

The architecture of the space instance demonstrator is a set of up to four identical (replicated) channels. Each channel is logically composed of two parts:

- The intra-channel part, in which the (replicated) application is executed.

- The inter-channel part, in charge of managing the channel dimension of the architecture. This management is done in a replicated distributed way.

At the intra-channel level, the application processing is segregated into two different CPU boards. The objective is to have one of the lanes (the secondary lane) act as a back-up for the other lane (the primary lane). Each lane supports a different operating system and different application software;

- The primary lane runs a full-functionality version of VxWorks and a nominal application that provides full control of the spacecraft and its payload. The application includes built-in self-monitoring based on executable assertions and timing checks.
- The secondary lane runs a much simpler, restricted version of VxWorks and a safety-monitoring and simple back-up application. The purpose of the latter is to provide control of the spacecraft in a very limited "survival" mode (e.g., sun-pointing and telemetry/telecontrol functions).

The idea is that neither the full VxWorks nor the nominal application supported by the primary lane can be trusted to be free of design faults. However, the restricted version of VxWorks and the application software supported by the back-up lane are assumed to be free of design faults and thus trustable. The aim is to allow continued (but severely degraded) operation in the face of a correlated fault across all processors of the primary lane. Errors due to such a correlated fault can be detected in two ways:

- By the self-monitoring functions (essentially control-flow monitoring and executable assertions) included within the nominal application (on the primary lane).
- By a safety-monitoring application executed by the secondary lane while the primary lane is operational.

The demonstrator supports two levels of integrity, corresponding to the differing levels of trust of the applications supported by the primary and secondary lanes. The nominal application (on the primary lane) is not trusted, so it is assigned to the lower integrity level. The back-up application is assumed to be free of design faults and is placed at the higher integrity level. This separation of the integrity levels on different lanes provides improved segregation (firewalling) between the two levels.

10.2.2 Physical Architecture

The space instance demonstrator is composed of up to four channels. Each channel is physically composed of two racks, one standard VME rack and one proprietary exclusion logic rack from Ansaldo Segnalamento Ferroviario (ASF) (Figure 10.5).

The standard VME rack contains the following boards:

- One nominal host board (FORCE 5CE, under full VxWorks), which executes the nominal application software and the inter-channel management software (pool manager).

- One backup host board (FORCE 5CE, under minimal VxWorks), which executes the safety monitoring software and also contains the backup software.
- One ICN manager board (MVME162-212, including two IP-COM360 piggy-back boards), which executes the ICN software

The proprietary ASF rack contains the following boards:

- Two ICN interface boards (ICN-I).
- Three PDC-k boards.
- One or two ANFS boards[1].

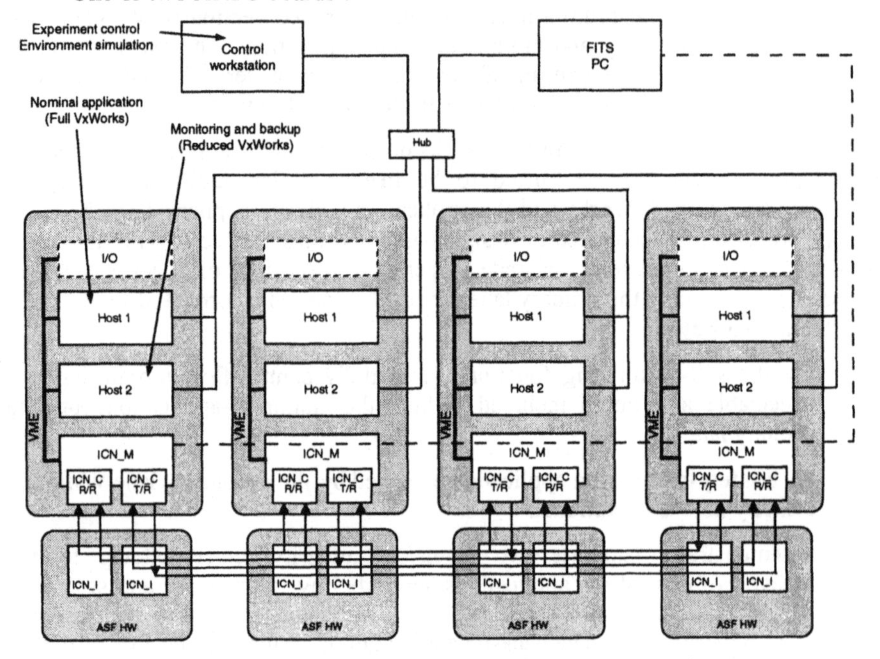

Figure 10.5 — Space Demonstrator Hardware Architecture

Furthermore, the GUARDS space instance demonstrator is connected to two external computers:

1. A workstation, in charge of executing the environment simulation software, and the man-machine interface of the demonstration. This workstation is connected to all host boards through a dedicated Ethernet I/O network.
2. A personal computer, implementing the FITS tool, connected both to the ICN boards through serial lines, and to host boards through the Ethernet I/O network.

[1] All channels have one ANFS board with a switcher; channels 3 and 4 also contain an ANFS board without a switcher.

10.2.3 Static Architecture

The static architecture (Figure 10.6) describes the software architecture from the viewpoint of its composing objects in terms of source files, procedures, packages, etc.

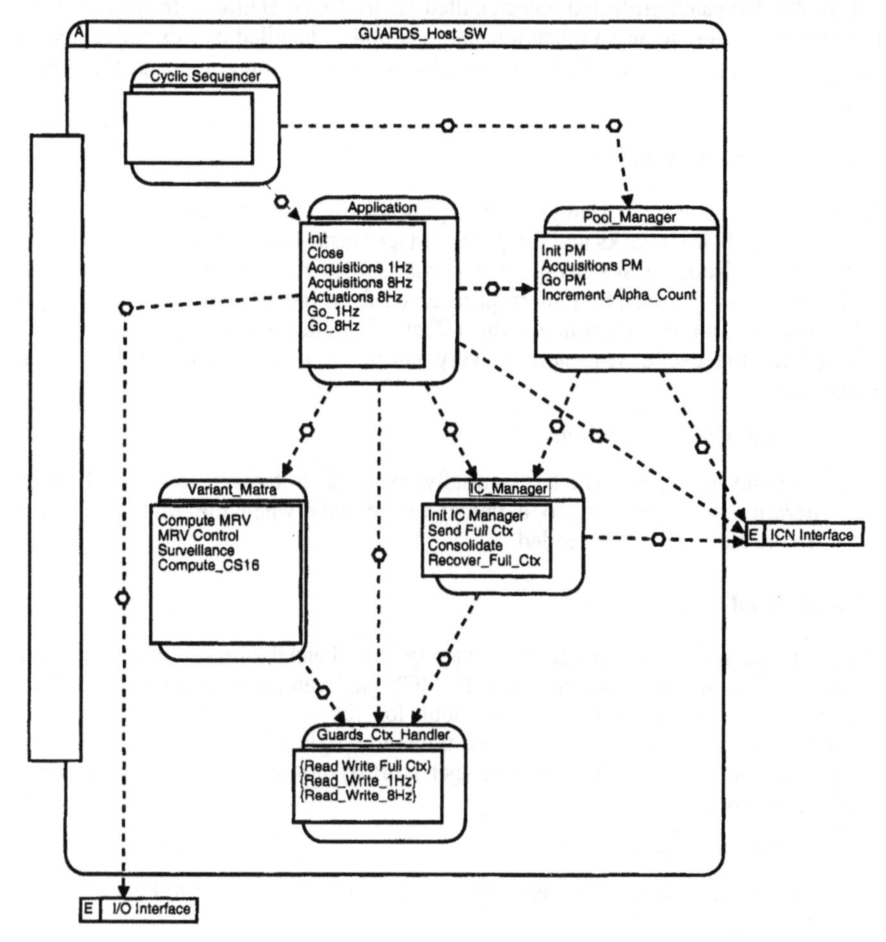

Figure 10.6 — Static Software Architecture

The main objects of the static architecture are:

- The cyclic sequencer.
- The pool manager.
- The application package, consisting of three objects, called: Application, Variant_Matra, and IC_Manager.
- The I/O interface library.
- The ICN interface library.

Each of these objects provides an initialisation operation, to be called at startup (or at restart), and involves one (or several) task(s), except for the ICN library interface.

An important point to note is that the full application context is under the control of a dedicated protected object, called Guards_Ctx_Handler. In other words, all read/write access to any application state variable is handled by this object. Such a design is necessary to allow a tractable channel reintegration process (see Chapter 4).

10.2.3.1 Cyclic Sequencer

The cyclic sequencer is in charge of managing all the cyclic tasks of the software, such as the application tasks and the pool manager task. Its function is to cyclically release the application tasks inside each frame, and to resynchronise the local host board with the synchronisation interrupt received from the ICN at the beginning of each frame (by tuning accordingly the offset of application tasks). The interface provided for the cyclic sequencer is very simple — it provides just two main operations:

- An initialisation function.
- A main function, which essentially executes a forever loop (each loop iteration corresponding to a frame) exited only when the mission of the application software is ended.

10.2.3.2 Pool Manager

The pool manager is in charge of managing the channel dimension. It takes as inputs the ICN internals coming from the ICN software, and elaborates the ICN control vectors to drive the ICN behaviour for the next cycle. It manages the host/ICN states according to a predefined transition logic. Like the cyclic sequencer, the interface provided for the pool manager is also very simple: it provides three main operations:

- An initialisation function.
- Two main functions, sequentially called at each cycle, and implementing the pool manager logic.

10.2.3.3 Application Package

The nominal application of the demonstrator is an MRV[2] mode management software, whose main mission is to reduce the rotational speed of the satellite after its separation from the launcher (Figure 10.7).

This application software is essentially composed of one object (itself internally decomposed) that provides the following entry points:

- *Acquisitions_1HZ*, which reads gyro increments and status.

[2] MRV stands for "Mode de Réduction de Vitesse" (speed reduction mode).

- *Processing_1HZ*, which calls sequentially *Compute_MRV* and *MRV_Control*.
- *Acquisitions_8HZ*, which reads the residual return current (EV current) from the thrusters to check that the appropriate thruster was activated.
- *Processing_8HZ*, which calls sequentially *Surveillance* and *Nozzle Modulation*..
- *Actuations_8HZ*, which sends thruster orders.

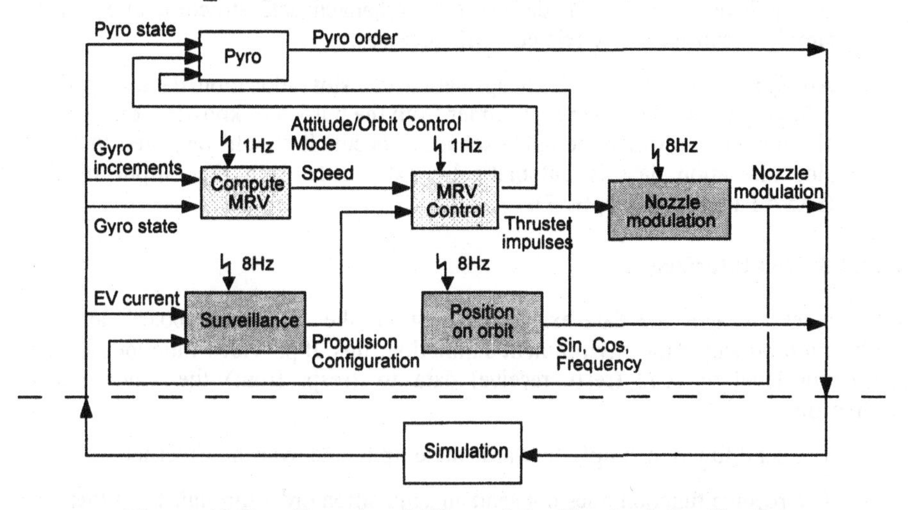

Figure 10.7 — Functional View of the Nominal Application

The application software uses two interface libraries:

- The ICN interface to read/write information from/to the ICN.
- The I/O interface to send/receive data to/from the environment simulator.

The application object is internally decomposed into three objects. The first one (Application) structures the application in terms of GUARDS transactions. The second object (Variant_Matra) contains the functional application itself. The third one (IC_Manager) manages interactive consistency and voting. In particular, IC_Manager provides a "Consolidate" operation to application transactions: in case of a detected discrepancy, the α-count of the corresponding channel is incremented.

The activity of porting this existing application onto the GUARDS architecture implied the following steps:

- Adaptation of the application real-time design as a set of ICN transactions.
- Modification of the I/O management, to cope with the simulated I/Os.
- Development of the interactive consistency logic to be applied on input values.

- Development of the result consolidation logic (by majority voting) to be applied on computed output values. In case of a channel identified as faulty, its α-count is incremented through a request towards the pool manager.

10.2.3.4 ICN Interface

This object encapsulates all the interface logic towards the ICN DPRAM. It is internally decomposed into two layers:

- A low-level layer, which defines the exchanged data structure and which simply encapsulates the related VME accesses.
- An application-level layer, encapsulating the slot allocation data, and thus relieving the application programmer from the burden of knowing the precise slot numbers in which the ICN exchanges occur. It should be noted that the slot allocation data is automatically extractable from the output of the GASAT tool (see Chapter 7).

10.2.3.5 I/O Interface

This object manages the data exchanges between the local host board and the workstation on which the environment simulation runs. It provides functions to the application-level to send (resp. receive) data to (resp. from) the environment simulation:

- The send function simply forwards the output data over the Ethernet.
- The receive function does not send an acquisition order through the Ethernet, but only returns a locally stored value. In fact, there is an internal handler which receives and stores the input data that is autonomously and cyclically transmitted by the simulation workstation.

10.2.4 Dynamic Architecture

The dynamic architecture (Figure 10.8) describes the software architecture from the viewpoint of its composing objects in terms of tasks, synchronisation resources, etc.

10.2.4.1 HRT Principles

The software executed on the host board must obey the HRT principles. We recall here very briefly the two main consequences of these principles:

1. The only active objects (i.e., tasks) which are allowed are either cyclic or sporadic.
2. The only way to communicate between two active objects is to use either an asynchronous communication interface (e.g., message passing) or to use an intermediate protected object, implementing the immediate priority ceiling (IPCI) protocol.

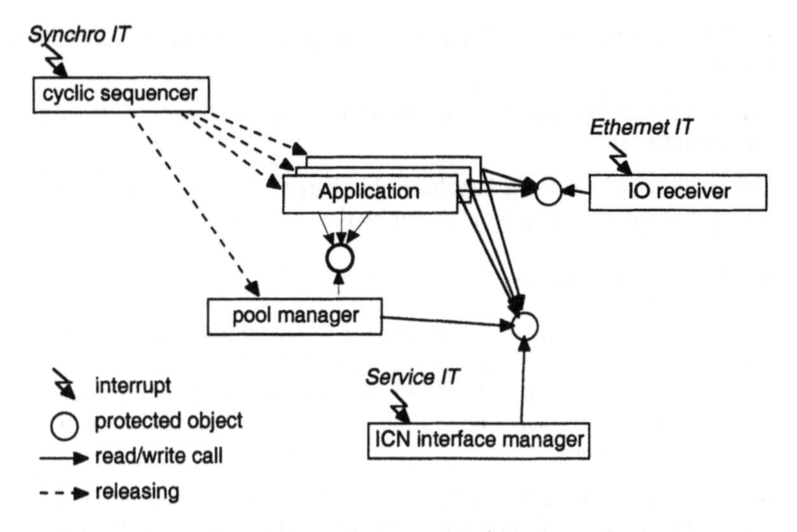

Figure 10.8 — Dynamic Architecture

10.2.4.2 Task Overview

The software running on the host board includes the following tasks:

- The cyclic sequencer, in charge of releasing all cyclic tasks.
- The 2 tasks of the pool manager transaction, in charge of managing the channel dimension.
- The 3 tasks of the 8 Hz application transaction.
- The 2 tasks of the 1 Hz application transaction.
- The I/O receiver, in charge of receiving data transmitted by the environment simulator.
- The ICN interface manager, in charge of updating the internal state of the ICN interface library.

From a real-time viewpoint:

- The cyclic manager task is cyclically released by the reception of the VME synchronisation interrupt from the ICN board.
- The application tasks and the pool manager tasks are all released by the cyclic sequencer.
- The I/O receiver performs a waiting receive on Ethernet inside a forever loop.
- The ICN interface manager waits on a semaphore (released by the VME interrupt service) inside a forever loop.

There are also some interrupt handlers in the dynamic architecture (cf. Figure 10.8):

- The VME service interrupt handler is in charge of releasing the ICN interface manager.
- The VME synchronisation interrupt handler is in charge of releasing the cyclic sequencer.
- The application watchdog handler is in charge of releasing the cyclic tasks (the associated watchdog is scheduler by the cyclic sequencer).

10.2.4.3 Nominal ICN Table

The nominal ICN Table is a frame lasting 1 second, and is decomposed into 8 resynchronisation cycles of 125 milliseconds each. Table 10.1 gives an overview of the different computations executed on the host board and on the ICN board during one resynchronisation cycle, according to the current slot. The slot number has the following precise meaning:

- When applied to a host computation, the slot number corresponds to the release time of the corresponding cyclic task. Of course, the 1Hz task is only released once per frame. On the contrary, the 8Hz task and pool manager tasks are released at each resynchronisation cycle.
- When applied to an ICN computation, the slot number corresponds to the physical DPRAM slot to be read or written.

Table 10.1 — Nominal ICN Table

Slot	Host Computation	ICN Computation
1	Sequencer	-
8	-	Icn_Internal directive
20	Acquisitions Pool Manager	-
30	Acquisitions 8Hz Task	-
40	Acquisitions 1Hz Task	-
53	-	IC 8Hz data
63	-	IC 1Hz data
70	Processing 8Hz Task	-
73	-	IC Pool Manager data
92	-	Vote 8Hz output data
100	Actuation 8Hz	-
110	Processing Pool Manager	-
120	-	Ctrl_Vector directive

10.2.4.4 Bootstrap/Joining Sequence

All objects of the architecture provide an initialisation function, which execute a number of preliminary initialisations, like creating some semaphores, allocating some global memory areas, etc. More precisely, the following initialisation sequence is performed at bootstrap:

- init_io_manager.
- init_icn_interface.
- init_cyclic_sequencer.
- init_application.
- init_pool_manager.

An important point to note is that the last step of the sequence is the initialisation of the pool manager, at the end of which the start command is sent towards the ICN software. Then, after a certain interval, the cyclic sequencer will receive a synchronisation interrupt and all the cyclic software starts to execute. At this point, two situations may arise:

1. The bootstrap sequence corresponds to the local channel executing a join. In this case, the pool manager restores the application context, and then starts the local application on the next frame.
2. The bootstrap sequence corresponds to the initial start of the whole demonstrator. In this case, once all channels are recorded as synchronised, the application starts executing from its initial default context (statically defined in the source code).

* * *

Figure 10.9 shows a photograph of the space demonstrator prototype. Here we can see two channels in the top half of the bay, and two channel in the bottom half. Each channel is composed of one ASF rack (horizontal racks) containing the proprietary hardware boards implementing the ICN and switch-off functions, and one standard VME rack (vertical racks) containing both the host processor board and the ICN_Manager board (which is connected to the corresponding ASF rack through a large black flat cable).

10.3 Railway Demonstrator

The application software of the railway demonstrator is derived from an existing proprietary control system. We first describe this proprietary system (an onboard control system for the new generation of trains for the Italian railroad system) and then the derived GUARDS demonstrator.

10.3.1 Railway Application System Architecture

The system has significant dependability requirements in terms of availability and predictability. To meet these requirements, it exploits state-of-the-art solutions both in its hardware and software architecture, such as a replicated communication bus, duplicated processing nodes and replicated subsystems. The system can be considered as representative of a wide class of real-time, embedded dependable systems.

Figure 10.9 — Space Demonstrator

The basic system architecture is composed of a number of subsystems connected through a PROFIBUS communication bus. Figure 10.10 illustrates the basic overall architecture of the system, showing the PROFIBUS and the various subsystems that communicate over it. Each subsystem plays a specific role in the overall real-time control of the train.

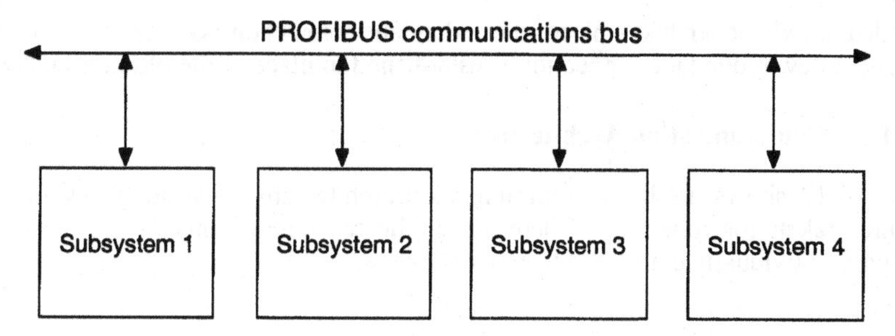

Figure 10.10 — Overall System Architecture

Each subsystem can be replicated for fault-tolerance. The replicated subsystems are in so-called *hot redundancy*, i.e., they maintain the same status, performing the same operations at all times, but the outputs of one replica are disabled. The exchange of messages on the bus is a critical factor in the co-operation of the subsystems to implement the overall system functions. As a consequence, message distribution mechanisms have very high availability and safety requirements. The basic system architecture choices to match these requirements are:

- The replication of the Profibus bus, as illustrated in Figure 10.11.
- The use of a special protocol.
- The elaboration in parallel of the messages in the components of a 2/2 subsystem architecture.

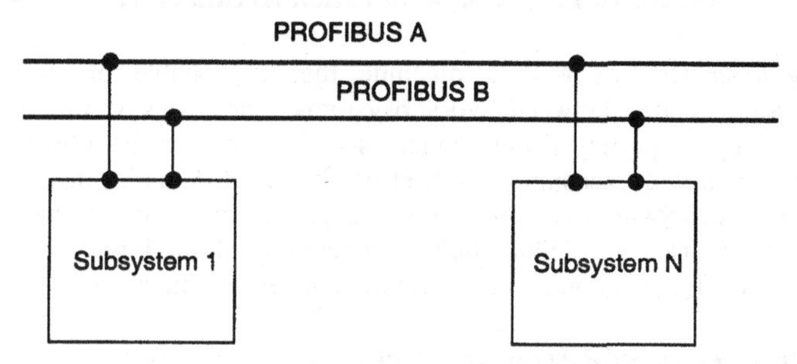

Figure 10.11 — Redundant Bus Architecture

The PROFIBUS protocol is connection-oriented. There are three types of messages:

- Control (connect, accept, switchover, disconnect, etc.).
- Data.
- Life (periodic message always sent on both busses in absence of other messages within a specified timeout period).

Two subsystems that need to interact with each other must establish a double connection: a *nominal* connection along which Control, Data and Life messages are exchanged, and a *redundant* connection where only Control and Life messages are exchanged. When a predefined threshold of errors on the nominal connection is exceeded, a switchover takes place: the redundant connection becomes the nominal one, and a new redundant connection is established in place of the old nominal one.

10.3.1.1 Communication Architecture

Figure 10.12 shows the basic communication architecture, illustrating several key measures taken for reliability. There are duplicated components (a 2/2 scheme), which are individually connected to separate buses.

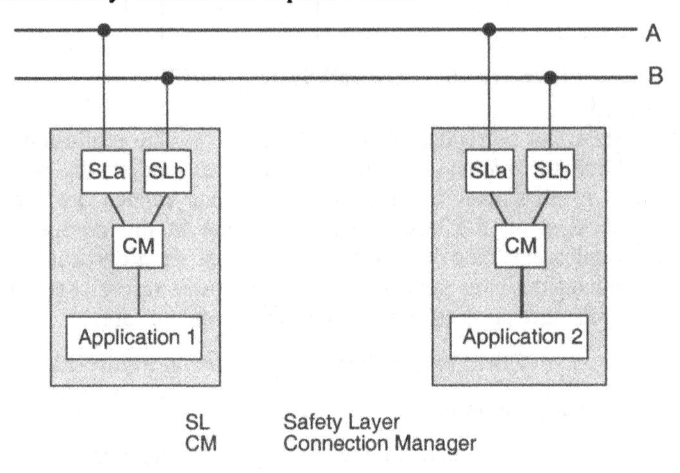

Figure 10.12 — Communication Architecture

A *safety layer* (SL) is a state machine that is capable of initiating and maintaining a connection between two subsystems. Each subsystem contains two safety layers. Each safety layer communicates with a peer in the other subsystem over a separate bus and exchanges a subset of the Control and Life messages. A safety layer receives Switchover messages and passes them up to the *connection manager* (CM), which controls the double connection. All the logic associated with the Switchover message is contained within the connection manager.

10.3.1.2 Proprietary Subsystem Architecture

Figure 10.12 would be valid even for a single (non-duplicated) physical architecture. In Figure 10.13, we illustrate with more precision the layout of the duplicated "two-out-of-two" (2/2) architecture.

There are two identical boards connected by various communication lines, and the two CPUs run identical software. There are hardware and software mechanisms for verifying that the computations carried out by each board are consistent.

The 2/2 architecture has two important consequences:

1. Each board is connected to only one PROFIBUS; in principle, each CPU does not have visibility of the messages of the other. This means that each message must be distributed to the other CPU in order to process it in parallel.

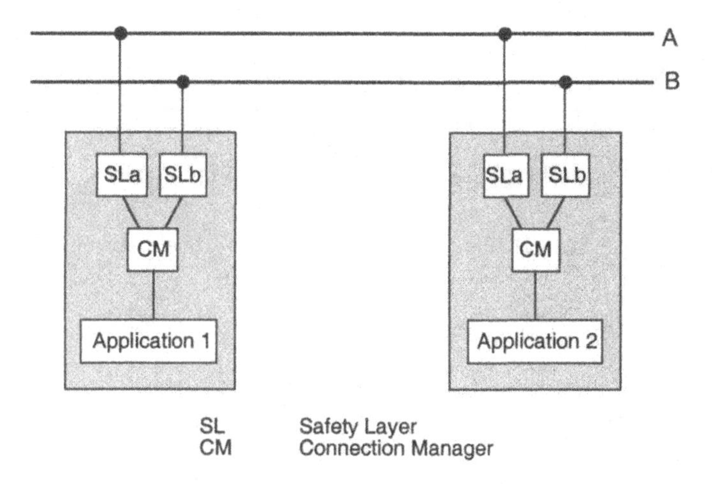

Figure 10.13 — Proprietary Subsystem Architecture

2. Message ordering must be the same in both CPUs, and therefore an additional mechanism is needed to buffer both input and output messages, and to agree upon the next message to process.

10.3.1.3 Bus Module Architecture

The additional mechanisms described in the second point mentioned in the previous section are implemented in the Event Handler that is in charge of controlling the 2/2 architecture, as shown in Figure 10.14. The Event Handler manages all interactions with the actual PROFIBUSes, and feeds its respective state machine with a consistent sequence of events.

The watchdog is not replicated (and therefore is a single point of failure—in fact, it is very expensive because it cannot be allowed to fail). At regular intervals, each of the (replicated) software systems must send a series of signals to the watchdog. If either of the software systems should miss a deadline, the watchdog disables both of them.

10.3.2 GUARDS Demonstrator System Architecture

In the GUARDS demonstrator version of the on-board system, a two-channel instance is used, with one channel for μA, and one for μB, as depicted in Figure 10.15. The inter-channel communication network carries the same data flow as that supported by the IPC connections in the original architecture. The Watch Dog is replaced by the GUARDS Exclusion Logic. Such a GUARDS instance is functionally equivalent to the proprietary 2/2 subsystem.

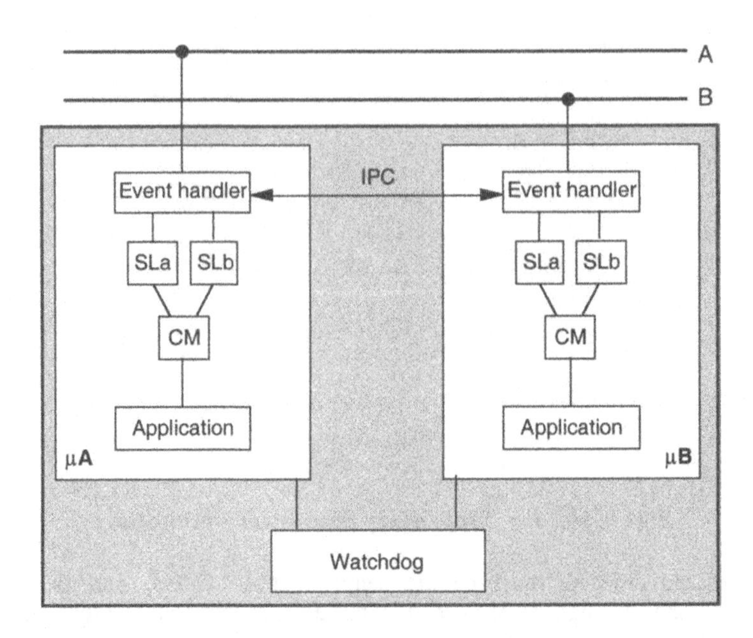

Figure 10.14 — Bus Module Architecture

10.3.3 Railway Application Software Architecture

In the existing railway application software, the communications manager kernel described above is part of the module that includes the software components reusable on each subsystem. The software layers are depicted in Figure 10.16, where:

- Application software contains the railway signalling logic. This layer is not aware of the 2/2 architecture.
- Specific device software contains activities that are different from subsystem to subsystem, or that are specific to each subsystem. They are aware of the 2/2 architecture and vote, when needed, on the results of their computation.
- Common software contains those parts of the basic services that must be implemented on all subsystems and do not differ from subsystem to subsystem.

The demonstrator includes a (substantial) part of the common software and modules representative of the specific device software and application software. All these software modules are seen as "application software" from the point of view of the GUARDS mechanisms.

The current demonstrator does not include the management of multiple levels of integrity.

* * *

Figure 10.17 shows a photograph of the railway demonstrator prototype.

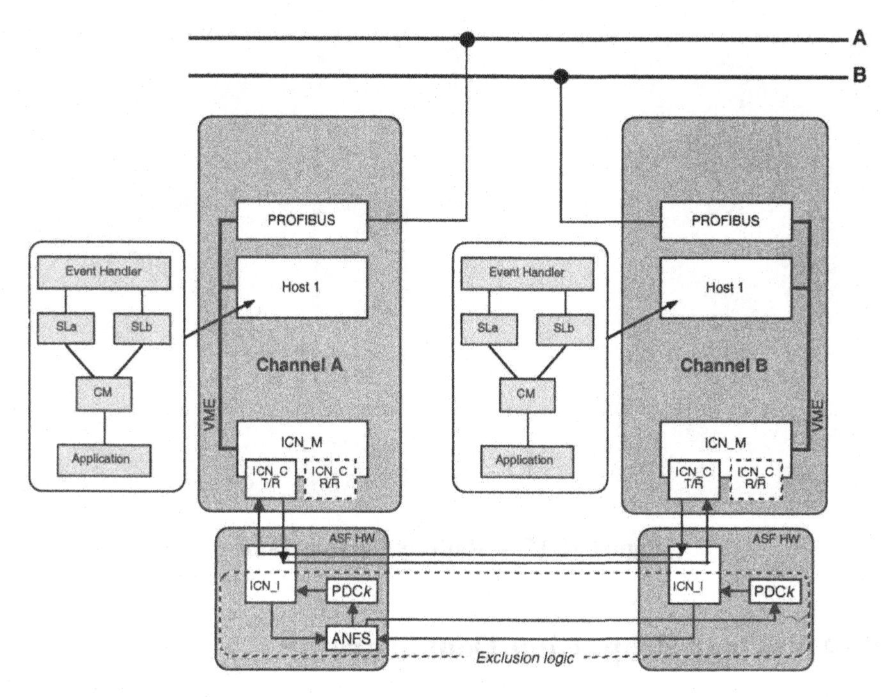

Figure 10.15 — Railway Demonstrator Architecture

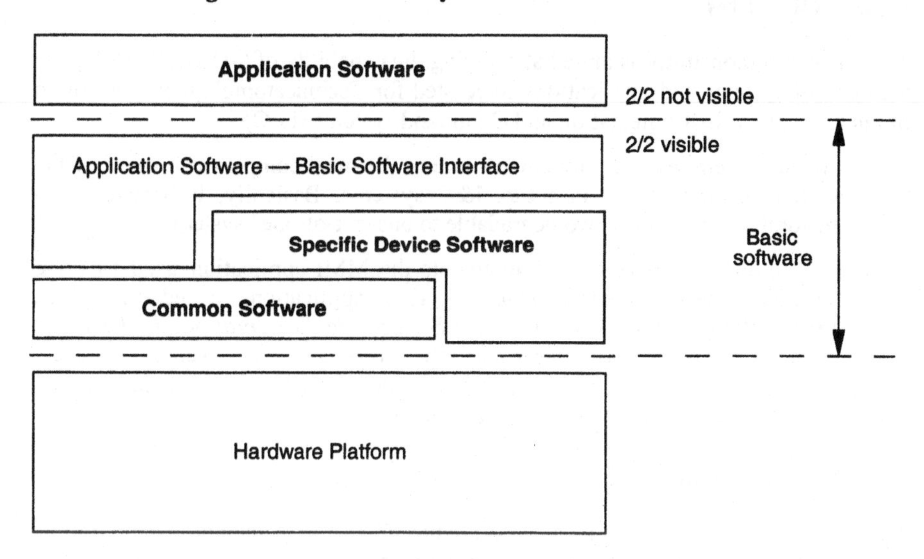

Figure 10.16 — Software Layers

Figure 10.17 — Railway Demonstrator

10.4 Nuclear Propulsion Demonstrator

10.4.1 Objectives

The nuclear demonstrator is aimed at showing the capability of GUARDS to handle the class of systems and applications forecasted for Technicatome's next generation of nuclear reactor Instrumentation and Command systems (I&C):

- At the system level, it implements one of the most complex use cases of the GUARDS architecture in future I&C systems. Basically, it behaves as a restorable "two-out-of-two degradable to one-out-of-one" system.
- At the application level, it implements the MMI application used to drive Technicatome's embedded reactors. This application, called the "Data Presentation System" (or SPD, for *Système de Présentation de Données*) acquires information concerning the process state (e.g., pressure, temperatures, alarms, etc.) via a field bus, and digital and analog inputs, and elaborates a view of the system state for a human operator.

10.4.2 Dependability Architecture

Technicatome's instance of the generic architecture is a 2-channel, 2-lane and 1-integrity level computer. The behaviour in the presence of faults is the following:

- Nominally, the instance works in a 2/2 configuration.
- Should an error be detected, it switches:
- to a 1/1 configuration if the error is detected within a single channel,

- to the safe state if the error is detected as a discrepancy between the two channels.

With respect to a classical 2/2 system, which is immediately put in the safe state once an error is detected, this strategy enhances availability without requiring a third channel[3], but at the price of a slight degradation of safety—since the system will work in a 1/1 configuration while the faulty channel is repaired.

In the demonstrator, both lanes of each channel use the same operating system (QNX). However, the architecture does support the diversification of the OS in one channel (one lane using QNX and the other one using VxWorks, for example). The two channels execute strictly the same code and the two lanes execute basically[4] the same code.

10.4.3 Hardware Architecture

The two channels are identical; they are both composed of one proprietary rack containing ASF's proprietary hardware and one standard VME rack.

The proprietary rack contains:

- The ICN Ethernet coupler (i.e., ICN-I).
- One PDCK board.
- One ANFS board (in one channel only).

The standard VME rack contains:

- One Motorola MVME-162 CPU board (ICN-M) with one IPCOMM 360 piggy back (ICN-C).
- Two SBS-OR VC5 Pentium 200 PC boards with 32 Mb of RAM and 8Mb of flash memory (host boards),.
- One M2I M-module FIP coupler (1 Mbit/s, single medium).
- An interface board allowing the fault injection tool (FITS) to reset the rack.
- A distributed hardware voter (the RMB).
- One SBS-OR VMIO digital input/output board (32 digital inputs, 16 relay outputs) and one SBS-OR VMIO analog input/output board (16 analog inputs, 4 analog outputs).

With the exception of the RMB board, the second rack is exclusively composed of COTS items.

---

[3] This architecture also maps directly onto the two sides of the architecture of the I&C system.

[4] The lanes are not quite symmetric due to a limitation of the current ICN in that only a single communicating entity is supported.

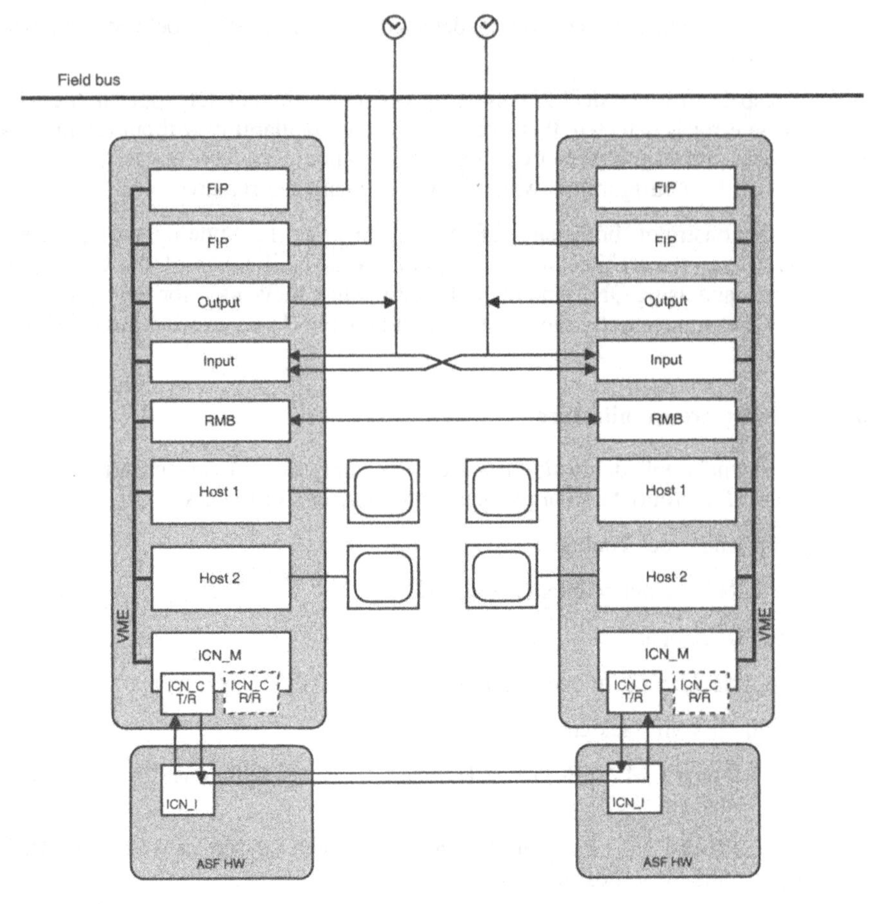

Figure 10.18 — Hardware Architecture

10.4.4 Software Architecture

The overall software architecture is depicted on Figure 10.19. The components involved in the architectureare not detailed in this book[5], focus is rather put on the processes implemented by these components (see Section 10.4.6 below).

10.4.5 Communication Architecture

There are three links between the channels:

- The ICN network, used for data exchange (including the interactive consistency protocol), voting (bit-to-bit comparison) and channel synchronisation.

[5] More details may be found in [Jenn 1999].

- The readback lines, used to propagate the digital and analog outputs between channels; these lines are use to perform the output cross-checking.
- The RMB shared states, used to perform the ultimate distributed hardware vote.

Inside each channel, the two lanes communicate via a shared memory region located in the ICN board. This link is essentially used to perform the cross-checking of data and control flows between hosts.

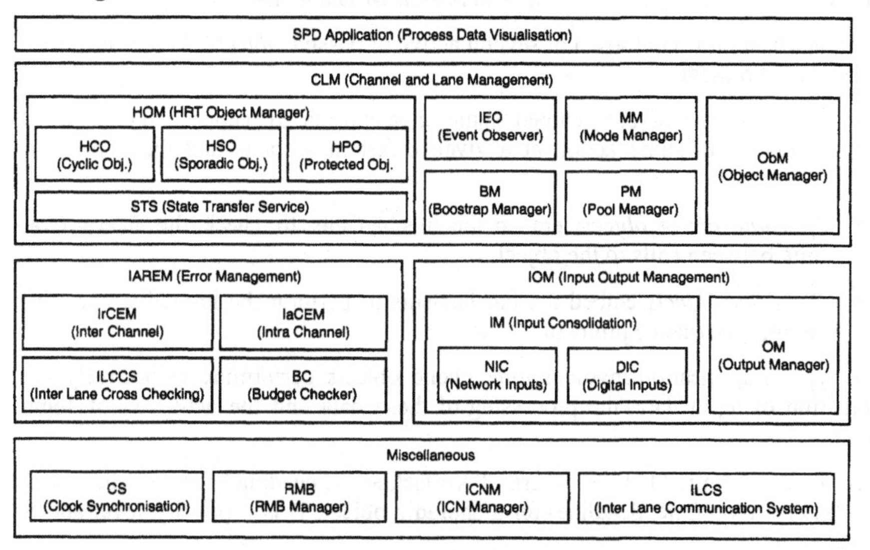

Figure 10.19 — Software Architecture

10.4.6 The Processes

This section describes the main processes[6] involved in the management of the instance:

- *Configuration management* process.
- *Communication* processes.
- *Pool management* process.
- *Fault-tolerance* processes.
- *Start-up* process.
- *Reintegration* and *state restoration* processes.
- *Input and output consolidation* processes.

---

6 Here, the word "processes" does not refer to a flow of control, but to an abstract sequence of activities.

10.4.6.1 Configuration Management Process

The management of the nuclear instance is organised around the concept of a *configuration.* At any one time, a channel is in a given configuration, corresponding to a given phase in the life-time of a phased-mission system or to a given phase in the life-time of the channel (e.g., start-up phase, reintegration, etc.).

A configuration is a tuple $C=(F, \{(O_1, M_1),..., (O_i,M_i), ..., (O_n,M_n)\})$ where F is an *ICN frame* (see § 10.1.2.2, O_i is an *object,* M_i is a *mode.*

An application is composed of several active or passive *objects.* Three categories of objects are provided:

- *Cyclic active objects* embed some mechanisms to support the periodic activation of some action at a given offset in a frame and the control of deadlines.
- *Sporadic active objects* embed the mechanisms to check the inter-arrival time between calls to the object.
- *Protected objects* embed the mechanisms to perform the immediate priority ceiling inheritance protocol.

A typical application involves three cyclic objects performing respectively the acquisition of input data, the processing of these data and the actuation of some device.

A *mode* is a tuple (T, P, S) where: T=(offset,period,deadline) for a cyclic object or T=(inter-arrival time, deadline) for a sporadic object; P is a priority level and S is a *selector.*

The *selector* is used to choose among several actions to be performed when the object is activated. These actions may corresponds to various functional needs, to different performance levels (e.g., nominal, degraded) or to different "roles" with respect to the overall state of the channel. In particular, this selector can be used to distinguish an application whose current role is to restore its internal state (i.e., a *learner*) from an application whose current role is to provide its internal state (i.e., a *teacher*).[7]

The configuration of a channel may change with time. This is typically the case when the channel starts up, leaves or reintegrates the pool, but it may also be the case for purely functional needs.

A configuration covers *two* dimensions: the set of objects and the modes of these objects. Consequently, changes may affect one or the other, or both of these dimensions. For example, during a state reintegration phase, the set of active objects may change (e.g., the *non-vital*[8] objects are eliminated from the active

---

7 In this prototype, state recovery is implemented as an extension of each application object, instead of by generic sweeper and catcher tasks (cf. Chapter 4), since QNX does not support multithreading.

8 *Vital* objects are objects that cannot be stopped at any time during the life-time of the system. If these objects are stopped, the system is in the safe state, and reciprocally. On the contrary, *non-vital* objects may be stopped at certain times, for example, to decrease the CPU or ICN load.

configuration and a state checker object is activated). Furthermore, the modes of the previously active objects may also change (from the "nominal" state to the "teaching" state, for example).

Configuration changes define the behaviour of the instance. Valid paths in the configurations space are defined by a state-transition graph. Configuration management essentially involves the *Object Manager* module (OM) and the *Mode Manager* module (MM).

10.4.6.2 Communication Processes

Three levels of communication are distinguished: (i) communication between the channels, (ii) communication between the lanes inside a channel and (iii) communication between the active objects inside a lane. The first level is described in detail in Chapter 2; consequently, focus is put on the remaining two communication channels.

Inter-host communication

The two lanes of a channel behave according to a strict 2-out-of-2 scheme. Therefore, exchanges between the two hosts are restricted to those necessary to perform cross-checking. They communicate via a shared memory area located (for the demonstrator) in the ICN memory space. To minimise coupling between the two host boards, inter-host communication relies on the "four slot fully asynchronous" algorithm, which guarantees (i) asynchrony of the two lanes and (ii) absence of race conditions.

The inter-host communication process mainly involves the *Inter Lane Communication* service (ILCS) for the data transfer and the *Inter Lane Cross-checking* service (ILCCS) for the cross-checking between the two hostsInter-object communication

Inter-object communication

Inter-object communication is exclusively based on shared memory. While this is not the natural way to communicate under QNX—which enforces the use of message-passing communication—it has been chosen for two main reasons: the absence of systematic synchronisation[9] and the standardisation of this API under POSIX — which will strongly facilitate OS-level diversification.

The inter-object communication process mainly involves the *Hard Real Time Objects* service (HRT).

10.4.6.3 Pool Management Process

The state of the pool of channels (in short, the *pool*) changes along time: a channel may be removed from the pool (isolated) or, conversely, may reintegrate the pool during start-up or after a maintenance operation. This redundancy management is called the pool management process.

---

[9] On the contrary, message passing—as used in QNX—imposes a systematic synchronisation between the sender and the receiver.

As for any other application, pool management is replicated on each channel. Each non-faulty channel maintains a consistent view of the pool so as to be able to take consistent decisions at the instance level concerning the evolution of the pool. To achieve this, each channel manages a finite state machine for each member of the pool—including itself—and makes its state change according to the occurrence of the *events* that may have an impact on the pool state. Examples of such events are: reception of a synchronisation frame from a channel trying to integrate or reintegrate the pool, error detection, reintegration authorisation, etc.

These events, which are collected by a specific object (the *Instance Event Observer*), are local opinions of a global event. Since the view of the pool must be identical on all non-faulty channels for consistent decisions to be taken, these local events must be consolidated at the inter-channel level before being used to fire the finite state machine transitions. The resulting *global* event is used to make the finite state machines of all channels change in a consistent manner. When a transition of a finite state machine is fired, an action is executed. This action may be to change the current configuration, to isolate a channel's output via the RMB, etc. Furthermore, the ICN configuration is systematically updated to reflect the pool state (enabling or disabling error detection, enabling or disabling the emission of the synchronisation message, etc.). Finally, the pool management process exports its view of the pool state to all objects which need this information; this concerns in particular the objects performing consolidation operations.

The pool management process mainly involves the *Instance Event Observer* (IEO) and the *Pool Management* (PM) services.

10.4.6.4 Fault Tolerance Processes

In the demonstration instance, the set of mechanisms related to error detection, error recovery and fault passivation is restricted to: (i) those exercising the specific features of GUARDS (ICN consolidation, instance reconfiguration, etc.) and (ii) those facilitating the debugging of the demonstration instance. For example, no watchdog has been implemented, even though this would certainly be one of the basic mechanisms to be included in an actual instance.

Intra-channel, intra-lane error detection

At this level, error detection is essentially based on the operating system error detection mechanisms and some consistency checks in a limited set of cases (e.g., when accessing cyclic buffers). In addition, some simple checks are performed to detect errors on acquisition data (out-of-range value, out-of-range dynamics).

Errors are handled via the C++ exception mechanism. The error-handling process is performed automatically and in a transparent manner with respect to the application code: the only explicit action consists in deriving new exception classes from the predefined classes and setting up the error-handling scheme. Basically, the error-handling sequence is the following:

- An exception is raised
- The error detection event is logged.
- The associated alpha-count mechanism is updated.

- If the alpha-count threshold is attained, the action of the error handler associated with the exception is executed. This action may be, for example, to propagate the error signalling at the inter-channel level or to check if the error is a common mode one, etc.

The error-handling process mainly involves the *Intra and Inter Channel Error Management* service (IarCEM)

Intra-channel, inter-lane error detection

The intra-channel inter-lane error detection mechanism is based on the comparison of *traces* elaborated by the two lanes. The traces cover the data produced by the two lanes (i.e., monitoring of the data flow) and the sequencing of the software (i.e., monitoring of the control flow). The traces are exchanged via the inter-lane communication service (see Section 10.4.6.2) which guarantees the absence of synchronisation points between the two hosts. This comparison process involves the *Intra-Channel Cross-Checking* module. It also involves an application-specific active object.

Inter-channel error detection

At the inter channel level, error detection is essentially carried out by the ICN, both for communication errors and for application errors (by means of the bit-to-bit vote directive).

Common-mode error detection

A specific process has been implemented to discriminate common-mode errors—occurring quasi-simultaneoulsy on several channels—from local errors. In the demonstration platform, this mechanism is used to identify and process appropriately common-mode faults coming from the instance environment (e.g., sensors). It is worth noting that the same mechanism could be used to detect common-mode faults affecting the software.

This process relies on the consolidation of error syndromes between the channels. Thus, when a channel detects an error that might be due to an external fault, error-handling is postponed until error consolidation has been carried out. The consolidation action consists in comparing the error syndromes and taking the appropriate action if the same syndromes have been observed by both channels.

This process involves the *Intra and Inter Channel Error Manager* service (IarCEM).

Inter-object error containment

This level of error containment has been addressed only very partially in our demonstration instance, the granularity level of passivation being the channel and not the active object. In particular, the issue of software of different levels of integrity running in a single host has not been investigated.

In the current implementation, inter-object error containment relies exclusively on the inter-process protection mechanisms of QNX, which themselves rely on the Pentium memory management unit (MMU).

Communication between active objects via shared memory constitutes a weak point with respect to error containment since it explicitly violates the confinement rules normally imposed by the MMU. Moreover, in this IPC scheme, the code that controls the shared data structure is replicated at each end of the IPC channel; consequently, an error may propagate from the application part of the active object to the shared data control code, leading potentially to an inconsistent state of the shared data structure. Typically, this could be the case for the shared cyclic buffers used by several GUARDS mechanisms: a faulty process may affect the top and bottom pointers in such a way that it will lead to the failure of the next clients of the shared cyclic buffer (even though it is not at the other end of the communication channel). In message-passing IPC, the shared cyclic buffer would be protected from such error propagation since the control data structure would only be modified by some code located in a specific process (a "server" process) which would then benefit from the MMU protection. However, the buffer control process would be a single point of failure.

Inter-lane error containment

From the external viewpoint, the two lanes perform strictly identical[10] operations. Consequently, data exchanges are limited to those required by the inter-lane cross-checking. An error in the data exchanged here will be detected by the inter-host cross-checking mechanism. However, an error in the control data structure of the communication buffer may propagate an error from one lane to the other (for the same reasons as those mentioned in the subsection on inter-object error containment). However, this would most likely lead to a detectable error and, consequently, to the passivation of the channel. Any error that remains undetected at this level will be detected by the inter-host cross-checking mechanism.

Inter-channel error containment

Containment of errors at the inter-channel level is ensured by: (i) the physical segregation of the channels, which reduces error propagation to a few well-identified paths, and (ii) by the consolidation of all inputs (including external inputs and data related to the management of the pool) by means of the interactive consistency protocol between channels.

Fault passivation

Three levels of fault passivation can be identified: (i) passivation of an active object, (ii) passivation of a channel and (iii) passivation of the complete instance.

The passivation of an active object may be done[11] by switching to a configuration that does not include the faulty object.

The passivation of a channel — or "suicide" — is done when the channel declares itself faulty, i.e., when the fault can be localised in one channel of the

---

[10] In fact, they are not perfectly symmetric in the current instance: one lane—the *master*—actually controls the ICN since the latter has not been designed to support more than one control host. The management of the input/output boards, which are shared by both hosts, is also asymmetric.

[11] This is not perfectly handled by the current software architecture since all object configurations and transitions from one configuration to another are predefined.

pool. The suicide consists of requesting the disconnection of the physical outputs via the redundancy management board and stopping the exchange of data on the ICN. It is important to note that the non-faulty channel must be informed of the intention of the faulty channel to commit suicide, otherwise the suicide would be considered as an inter-channel level error and, consequently, would lead to the complete passivation of the instance.

In a complementary manner, the passivation of the instance — or "isolation" — is done when the detected error cannot be ascribed to any particular channel. A communication error detected by the ICN is an example of such an "inter-channel level error". Indeed, such an error may be due to the failure of the sender or the receiver (e.g., failure of the Ethernet coupler), and a more accurate diagnosis must be carried out to decide which channel is actually faulty. In the demonstration instance, the scheme that has been chosen is the following: when an inter-channel error is detected, the instance is put in the safe state (both channels are isolated), both channels execute self-tests indefinitely until one of the two channels declares itself faulty and commits suicide[12]. Should this occur, the remaining channel is informed and reintegrates the pool alone.

The management of object level passivation is performed by the *Object Manager* and the *Mode Manager* services. The channel-level and instance-level passivation essentially involve the *Instance Event Observer* and the *Pool Manager* services (see Section 10.4.6.3).

Reconfiguration

Two reconfiguration policies have been retained depending on whether the error is detected at the intra-channel level or at the inter-channel level.

For an intra-channel level error (i.e., a fault can be localised in a particular channel), the channel which commits suicide resets itself. In the demonstration instance, this operation is considered to put the channel back in an error-free state. Note that a mechanism limits the number of successive "suicide-reset-pool reintegration" sequences: after *n* successful reintegrations, the channel is denied the right to reintegrate the pool. In practice, this means that it is definitely isolated.

For an inter-channel level error (i.e., a fault cannot be localised in a particular channel), both channels isolate themselves but do not perform reset. To restart the instance, some manual operation is required (e.g., error diagnosis and replacement of the faulty channel) and a full switch-on sequence is required (a more complex behaviour has been actually *partially* implemented, see the previous section).

10.4.7 Start-up Process

Globally, the start-up sequence is the following:

- The first channel is started (the two channels are symmetric and can be started in any order).

[12] Note that this is not a true suicide, that is to say a suicide involving the event observation and the pool management modules. This form of suicide only involve the RMB. Note also that this has not been fully implemented in the current instance.

- The first channel reaches its permanent state (i.e., its configuration will not change if no external events occur); it does not perform any actuation.
- The second channel is started (note that it must not be started before the previous condition is satisfied).
- The reception of the synchronisation messages by the two channels initiates the integration sequence.
- The two channels synchronise themselves at the ICN level.
- The first channel sends the relevant part of its internal state to the second channel.
- The two channels synchronise themselves at the application level.
- The two channels start executing their nominal actions.
- The outputs of both channels are connected to the actuators.

The start-up process is a specific case of channel reintegration; the difference lies in the fact that the first channel in the pool is not allowed to perform nominal operations (in particular, actuation) until the *quorum* is reached. On the contrary, during a reintegration phase, the remaining channel in the pool *does* perform nominal operations, even though it is alone.

In the 2-out-of-2 demonstration instance, the quorum is equal to two. This is an arbitrary choice. The other alternative, i.e., accepting the instance to start up with only one channel, would also be valid depending on the targeted dependability level. On a 2-out-of-3 instance, one could accept to start the instance with only 2 channels (i.e., the quorum is set to 2) since safety—but not availability—will be ensured during startup.

The concept of quorum is not hard-coded but corresponds to an appropriate set-up of the configuration management finite state machine (see Section 10.4.6.1).

10.4.8 Channel Reintegration and State Restoration Processes

A channel which has committed suicide is granted the right to reintegrate the pool once it has completed its reset sequence. This corresponds to the transition from a 1-out-of-1 configuration back to the nominal 2-out-of-2 configuration.

During a reintegration phase, the joining channel and the active channels execute neither the same ICN table nor the same configuration. Therefore, the reintegration process ensures the synchronisation of the channels: (i) on a per *cycle* basis (i.e., the clock synchronisation interrupts must occur synchronously on all channels), (ii) on a per *frame* basis (i.e., the ICN must execute the same frame on all channels), and (iii) on a per *configuration* basis (i.e., the same configuration must be executed by all channels).

The first synchronisation level is carried out by the ICN component, which executes the clock synchronisation algorithm. The host-based software has no control over this but has the capability to activate or deactivate the emission/reception of synchronisation messages via the ICN control vectors.

Achieving the second and third synchronisation levels requires some exchanges between the channels in order to communicate the current (nominal) configuration to the joining channel. This is done in two phases:

- First, the Mode Manager of the joining channel gets its own state from the other channels via the ICN Manager.
- Second, the Mode Manager of the joining channel sets up the channel application state so as to be in the same as the other channels.

The first phase requires some transaction-level synchronisation between the "teaching" channel and the "learning" channel since it involves an ICN exchange. This is ensured by the fact that the learning channel executes the Mode Manager each and every cycle (it executes table 0) and listens to the ICN at each cycle. When it receives some data in the expected slot, it considers that it is the Mode Manager state chunk.

Once the Mode Manager has received the (unique) state chunk, it sets up its internal state so as to be in the same configuration as the teaching channel. The state chunk contains essentially the current state of the Mode Manager protected object finite state machine and the "remaining cycles" information maintained by the teaching channel. Once the learning channel has received these data and overwritten its own values by the latter, it requires a mode change to occur at the next frame. This is also done in the teaching channel.

Since the remaining cycles value of the learning channel is the same as the one of the teaching channel, the actual mode change will occur at the same time.

At this time, the teaching and learning channel execute the *same* configuration, in terms of the instance management *and* application active objects. The application-level state restoration phase can now begin. This phase is mandatory for the two channels to behave in a deterministic manner so that it can be guaranteed that the two channels will produce the same outputs from the same inputs in the absence of faults. The state restoration is essentially based on configuration changes (cf. Section 10.4.6.1).

During the state restoration process, the channel currently in the pool (i.e., the *teacher*) and the channel willing to join the pool (i.e., the *learner*) execute different configurations:

- The teacher executes a nominal configuration (possibly with only the vital active objects), but the local modes of the active objects indicate that there is a "teaching" activity to be done along with the nominal activity. In the new configuration, the offset and the period of the vital active objects should (a priori) not change. Conversely, the non-vital objects that are kept active during the restoration phase may be activated with a greater period and/or lower priority than normally, so as to decrease the CPU load and compensate the load increase induced by the state transfer.

  During the state elaboration process, every teaching object adds a state chunk to its normal data flow towards the ICN (or uses a specific ICN slot dedicated to the state exchange) and compute a local "state digest". When the state transfer is completed (i.e., when the last state chunk has been sent), every active object sends the state digest to a specific active object (i.e., the State Checker) in charge of checking that the state received by the joining channel is correct.

- The learner executes a configuration corresponding to the one executed by the teacher. With respect to the latter, the main differences are the following:
- The learner does not execute the nominal parts of the active objects but simply gets the state chunks sent by the teacher (whereas the latter executes both the nominal activity and the state chunking operation).
- The offsets of the learner's active objects are chosen so as to support the ICN transfer, i.e., they have to be activated after the ICN slot where the state chunk exchange is done.

  The learner gets the state chunks from the ICN and builds the state digests. These digests are all sent to the State Checker, which is scheduled to be activated once all active objects have performed their state exchange. At that time, the State Checker broadcasts the set of state digests to all other channels. Each state digest is associated to an identifier of the active object from which it originates: this allows all State Checkers to compare the state digests of their active (and vital) objects. Currently, the behaviour consists in rejecting a candidate channel for which at least one state digest does not match with one of the channels that are currently active.

Note that the learner currently receives its state chunks from a single but arbitrary channel of the pool (one in the nominal active state). This mechanism could be improved by distributing the state chunks broadcast among the set of active channels, so as to reduce the CPU load of the teaching channels and the ICN load. A preliminary implementation of this mechanism is already available in the demonstration instance.

Channel reintegration mainly involves the *Mode Manager* (MM, and, consequently, the *Object Manager—OM*) and the *Pool Manager (PM)* services. The state restoration process involves mainly the *Hard Real Time objects* (HRT) service for the handling of the teaching/learning modes. The elaboration/restoration of state chunks is not carried out by a specific service since it is tightly dependent on the object state. However, the general scheme is given in the *State Checker* service.

10.4.9 Input/Output Consolidation Process

The input consolidation process basically involves four phases: (i) sample the input, (ii) broadcast the inputs among the channels, (iii) reject the incorrect inputs and, finally, (iv) elaborate a consensual input. To minimise the risk of incorrect rejection and to increase the accuracy of the measure, the sampling operation performed on the redundant channels must be as synchronous as possible.

This sequence can be fully applied on analog and digital inputs, however, the second phase cannot be performed on networked inputs since the bandwidth of inter-channel network is much lower than the field bus network. Consequently, the consolidation process is performed on a digest of the sample data (here a network frame). In the absence of error, each channel acquires the same sequence of frames possibly with a certain desynchronisation (i.e., one channel gets a frame before the other) and calculates the "same" sequence of digests. The consolidation process then consists in maintaining a buffer of the sampled frames, and select the frame corresponding to the consolidated digest. The depth of the frame buffer directly depends on the redundancy level and on the tolerable frame errors[13]. This approach is applicable if the source of data is unique and if the integrity of the frames can be ensured, which is easily ensured by means of frame checksum.

The output consolidation process is detailed in Chapter 5.

* * *

Figure 10.20 shows a photograph of the nuclear propulsion demonstrator

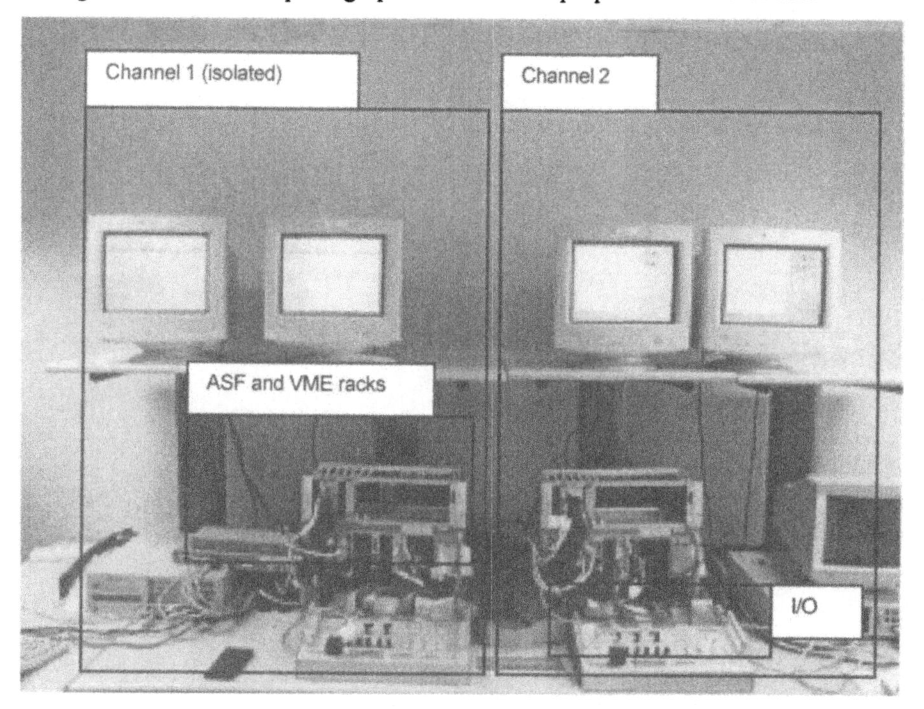

Figure 10.20 — Nuclear Propulsion Demonstrator

[13] Basically, for two redundant channels, in the absence of errors, a one-entry buffer is required.. Should a missing frame be tolerated, a supplementary entry shall be added.

Annex A

Project Consortium

- **Technicatome** *(coordinator)*
  1100, avenue Jean-René Guillibert
  Gautier de la Lauzière
  BP 34000
  F-13791 Aix-en-Provence Cedex 3
  France

- **Ansaldo Segnalamento Ferroviario**
  Via dei Pescatori 35
  I-16129 Genova
  Italy

- **Intecs Sistemi SpA**
  V. L. Gereschi, 32/34
  I-56127 Pisa
  Italy

- **LAAS-CNRS**
  7 avenue du Colonel Roche
  F-31077 Toulouse Cedex 4
  France

- **Matra Marconi Space France**[1]
  31 rue des Cosmonautes
  F-31402 Toulouse Cedex 4
  France

- **Pisa Dependable Computing Centre**

  PDCC-CNR (CNUCE and IEI)
  Area della Ricerca di Pisa
  Via A. Moruzzi 1
  Loc. S. Cataldo I-56100 Pisa
  Italy

  PDCC-Università di Pisa
  Dipartimento di Ingegneria
  dell'Informazione
  Via Diotisalvi 2
  I-56100 Pisa
  Italy

- **Siemens AG Österreich PSE**
  Gudrunstrasse 11
  A-1100 Wien
  Austria

- **Universität Ulm**[2]
  Fakultät für Informatik
  Postfach 4066
  D-89069 Ulm
  Germany

- **University of York**
  Department of Computer Science
  Heslington
  York YO10 5DD
  England

---

[1] Matra Marconi Space France became Astrium SAS in May 2000.
[2] Subcontractor to *Technicatome*.

Annex B
Abbreviations

ACTL	Action-based CTL
ADE	Architecture Development Environment
AMC	ACTL Model Checker
ANFS	Fail-safe AND boards (ANd Fail-Safe)
ANSI	American National Standards Institute
API	Application Programmer Interface
ASF	Ansaldo Segnalamento Ferroviario
ATV	Automated Transfer Vehicle
BA	Byzantine Agreement
BDD	Binary Decision Diagram
CASE	Computer Aided Software Engineering
CCS	Calculus of Communicating Systems
CM	Connection Manager
CNR	Consiglio Nazionale delle Ricerche
CNRS	Centre National de la Recherche Scientifique
CNUCE	Centro Nazionale Universitario di Calcolo Elettronico
COTS	Commercial Off The Shelf
CPU	Central Processor Unit
CRC	Cyclic Redundancy Check
CRV	Crew Return Vehicle
CTV	Crew Transport Vehicle
CTL	Computation Tree Logic

DPRAM	Dual-Port Random Access Memory
DSPN	Deterministic and Stochastic Petri Net
ESPRIT	European Strategic Programme for Research in Information Technology
FIFO	First-In First-Out
FIP	A field bus standard [EN 50170]
FITS	Fault Injection Tool-Set
FMECA	Failure Mode Effect and Criticality Analysis
GASAT	GUARDS Allocator Schedulability Analyser Tool
GSPN	Generalised Stochastic Petri Net
GUARDS	Generic Upgradable Architecture for Real-time Dependable Systems
HDE	HOOD Diagram Editor
HFR	Hazardous Failure Rate
HOOD	Hierarchical Object-Oriented Design
HRT	Hard Real Time
HRT-HOOD	Hard Real Time Hierarchical Object-Oriented Design
IC	Interactive Consistency
I&C	Instrumentation and Control
I/O	Input/Output
ICN	Inter-channel Communication Network
ICN_C	ICN Controller
ICN_I	ICN Interface
ICN_M	ICN Manager
IEC	International Electrotechnical Commission
IEI	Istituto di Elaborazione della Informazione
IEEE	Institute of Electrical and Electronic Engineers
IMA	Integrated Modular Avionics
ISO	International Organisation for Standardisation
IPC	Inter Process Communication
IPCI	Immediate Priority Ceiling Inheritance protocol
ISSE	Informal Solution Strategy Editor and Analyser
JACK	Just Another Concurrency Kit

LAAS	Laboratoire d'Analyse et d'Architecture des Systèmes
LTS	Labelled Transition System
MD5	A Message-Digest algorithm [Rivest 1992]
MLO	Multiple-Level Object
MMI	Man-Machine Interface
MMS	Matra Marconi Space France
MMU	Memory Management Unit
MRSPN	Markov Regenerative Stochastic Petri Net
MRV	Speed Reduction Mode (*Mode de Réduction de Vitesse*)
NP	Non-Polynomial
ODS	Object Description Skeleton
ODSE	Object Description Skeleton Editor
OS	Operating System
PDCC	Pisa Dependable Computing Center
PDCK	DC Power Supply n° K
PMS	Phased Mission Systems
POSIX	Portable Operating System Interface [IEEE 1003]
PTM	Phase Transition Model
PVS	Prototype Verification System
QNX	A commercial real-time operating system (QNX is a trademark of QNX Software Systems Ltd.)
RMB	Redundancy Management Board
RSA	A public-key cryptosystem, from Rivest, Shamir and Adleman, the authors of [Rivest *et al.* 1978].
SAN	Stochastic Activity Network
SIF	Standard Interface Format
SL	Safety Layer
SLO	Single-Level Object
SPD	Data Presentation System (*Système de Présentation des Données*)
SR	State Restoration
SWIFI	SoftWare-Implemented Fault Injection
TA	Technicatome

TMR	Triple Modular Redundancy
VME	VersaModule Eurocard bus
VxWorks	A commercial real-time operating system (VxWorks is a trademark of Wind River Systems)
WCET	Worst-Case Execution Time

References

[Adams 1989] S. J. Adams, "Hardware Assisted Recovery from Transient Errors in Redundant Processing Systems", in *19th IEEE Int. Symp. on Fault-Tolerance Computing (FTCS-19),* (Chicago, IL, USA), pp.512-9, IEEE Computer Society Press, 1989.

[Agrawal 1988] P. Agrawal, "Fault Tolerance in Multiprocessor Systems without Dedicated Redundancy", *IEEE Transactions on Computers*, C-37 (3), pp.358-62, 1988.

[Amendola & Marmo 1997] A. Amendola and P. Marmo, *Approach, Methodology and Tools for Evaluation by Analytical Modelling - Railway Application,* ESPRIT Project 20716 GUARDS Report N°D3A6.AO.1011.A, Ansaldo Segnalamento Ferroviario, Naples, Italy, March 1997.

[Anderson & Lee 1981] T. A. Anderson and P. A. Lee, *Fault Tolerance — Principles and Practice,* Prentice-Hall, 1981.

[Arlat 1997] J. Arlat, *Preliminary Definition of the GUARDS Validation Strategy*, ESPRIT Project 20716 GUARDS Report N°D3A1.A0.5002.C, LAAS-CNRS, Toulouse, France, January 1997 (LAAS Research Report N°96378).

[Arlat *et al.* 1978] J. Arlat, Y. Crouzet and C. Landrault, "Operationally Secure Microcomputers", in *1st Conference on Fault-Tolerant Systems and Diagnostics (FTSD'78),* (J. W. Laski, Ed.), (Gdansk, Poland), pp.1-9, 1978.

[Arlat *et al.* 1990] J. Arlat, M. Aguera, L. Amat, Y. Crouzet, J.-C. Fabre, J.-C. Laprie, E. Martins and D. Powell, "Fault Injection for Dependability Validation — A Methodology and Some Applications", *IEEE Transactions on Software Engineering,* 16 (2), pp.166-82, 1990.

[Arlat *et al.* 2000] J. Arlat, T. Jarboui, K. Kanoun and D. Powell, "Dependability Assessment of GUARDS Instances", in *IEEE International Computer Performance and Dependability Symposium (IPDS 2000),* (Chicago, IL, USA), pp.147-56, IEEE Computer Society Press, 2000.

[Audsley 1993] N. Audsley, *Flexible Scheduling for Hard Real-Time Systems,* D. Phil. Thesis, Dept. of Computer Science, University of York, UK, 1993.

[Audsley *et al.* 1993a] N. Audsley, A. Burns, M. Richardson, K. Tindell and A. J. Wellings, "Applying New Scheduling Theory to Static Priority Pre-emptive Scheduling", *Software Engineering Journal,* 8 (5), pp.284-92, 1993.

[Audsley *et al.* 1993b] N. Audsley, A. Burns and A. J. Wellings, "Deadline Monotonic Scheduling Theory and Application", *Control Engineering Practice*, 1 (1), pp.71-8, 1993.

[Audsley *et al.* 1993c] N. Audsley, K. Tindell and A. Burns, "The End of the Line for Static Cyclic Scheduling?", in *5th Euromicro Workshop on Real-Time Systems,* (Oulu, Finland), pp.36-41, IEEE Computer Society Press, 1993.

[Barrett *et al.* 1995] P. Barrett, A. Burns and A. J. Wellings, "Models of Replication for Safety Critical Hard Real-Time Systems", in *20th IFAC/IFIP Workshop on Real-Time Programming (WRTP'95),* (Fort Lauderdale, FL, USA), pp.181-8, Pergamon Press, 1995.

[Bate & Burns 1997] I. Bate and A. Burns, "Schedulability Analysis of Fixed Priority Real-Time Systems with Offsets", in *9th Euromicro Workshop on Real-Time Systems,* (Toledo, Spain), pp.153-60, IEEE Computer Society Press, 1997.

[Bate & Burns 1998] I. Bate and A. Burns, "Investigation of the Pessimism in Distributed Systems Timing Analysis", in *10th Euromicro Workshop on Real-Time Systems,* (Berlin, Germany), IEEE Computer Society Press, 1998.

[Béounes *et al.* 1993] C. Béounes, M. Aguera, J. Arlat, S. Bachmann, C. Bourdeau, J.-E. Doucet, K. Kanoun, J.-C. Laprie, S. Metge, J. Moreira de Souza, D. Powell and P. Spiesser, "SURF-2: A Program for Dependability Evaluation of Complex Hardware and Software Systems", in *23rd IEEE Int. Symp. on Fault-Tolerant Computing (FTCS-23),* (Toulouse, France), pp.668-73, IEEE Computer Society Press, 1993.

[Bernardeschi *et al.* 1994] C. Bernardeschi, A. Fantechi and L. Simoncini, "Formal Reasoning on Fault Coverage of Fault Tolerant Techniques: a Case Study", in *1st European Dependable Computing Conference (EDCC-1),* (Berlin, Germany), Lecture Notes on Computer Science, 852, pp.77-94, Springer-Verlag, Berlin, 1994.

[Bernardeschi *et al.* 1997] C. Bernardeschi, A. Fantechi and S. Gnesi, *Formal Specification and Verification of the Inter-Channel Consistency Network,* ESPRIT Project 20716 GUARDS Report N°D3A4.AO.6004.C, PDCC, Pisa, Italy, March 1997.

[Bernardeschi *et al.* 1998a] C. Bernardeschi, A. Fantechi, S. Gnesi and A. Santone, *Formal Specification and Verification of the Inter-Channel Consistency Network,* ESPRIT Project 20716 GUARDS Report N°I3A4.AO.6009.B, PDCC, Pisa, Italy, April 1998.

[Bernardeschi *et al.* 1998b] C. Bernardeschi, A. Fantechi, S. Gnesi and A. Santone, *Formal Specification and Verification of the Inter-Channel Fault Treatment Mechanism,* ESPRIT Project 20716 GUARDS Report N°I3A4.AO.6013.A, PDCC, Pisa, Italy, May 1998.

[Bernardeschi *et al.* 1998c] C. Bernardeschi, A. Fantechi, S. Gnesi and A. Santone, "Automated Verification of Fault Tolerance Mechanisms", in *3rd Int. Workshop on Formal Methods for Industrial Critical Systems,* (Amsterdam, Netherlands), pp.17-41, Springer-Verlag, 1998.

[Bernardeschi *et al.* 1999] C. Bernardeschi, A. Fantechi and S. Gnesi, "Formal Validation of the GUARDS Inter-consistency Mechanism", in *18th Int. Conf. on Computer Safety, Reliability and Security (SAFECOMP99),* (Toulouse, France), Lecture Notes in Computer Science, 1698, pp.420-30, PDCC, Pisa, Italy, 1999.

[Biba 1977] K. J. Biba, *Integrity Considerations for Secure Computer Systems,* Technical Report N°MTR-3153, Rev. 1, The Mitre Corporation, April 1977.

[Blough *et al.* 1992] D. M. Blough, G. F. Sullivan and G. M. Mason, "Intermittent Fault Diagnosis in Multiprocessor Systems", *IEEE Transactions on Computers,* 41 (11), pp.1430-41, 1992.

[Bondavalli *et al.* 1997a] A. Bondavalli, S. Chiaradonna, F. D. Giandomenico and F. Grandoni, "Discriminating Fault Rate and Persistency to Improve Fault Treatment", in *27th IEEE Int. Symp. on Fault-Tolerant Computing (FTCS-27),* (Seattle, WA), pp.354-62, IEEE Computer Society Press, 1997.

[Bondavalli *et al.* 1997b] A. Bondavalli, I. Mura and M. Nelli, "Analytical Modelling and Evaluation of Phased-mission Systems for Space Applications", in *2nd Workshop on High Assurance Systems Engineering (HASE-97),* (Bethseda, MD, USA), IEEE Computer Society Press, 1997.

[Bondavalli *et al.* 1997c] A. Bondavalli, S. Chiaradonna, F. D. Giandomenico and F. Grandoni, *Inter-Channel State Restoration,* ESPRIT Project 20716 GUARDS Report N°I1-SA4.TN.6006.B, PDCC, Pisa, Italy, November 1997.

[Bondavalli *et al.* 1998a] A. Bondavalli, S. Chiaradonna, F. D. Giandomenico and F. Grandoni, *Threshold-based Mechanisms to Discriminate Transient from Intermittent Faults,* Technical Report N°B4-17, IEI-CNR, Pisa, Italy, June 1998a.

[Bondavalli *et al.* 1998b] A. Bondavalli, I. Mura, X. Zang and K. S. Trivedi, *Dependability Modelling and Evaluation of Phased Mission Systems: a DSPN Approach*, ESPRIT Project 20716 GUARDS Report N°I3A4.AO.6010.A, PDCC, Pisa, Italy, January 1998b.

[Bondavalli *et al.* 1998c] A. Bondavalli, F. D. Giandomenico, F. Grandoni, D. Powell and C. Rabéjac, "State Restoration in a COTS-based N-Modular Architecture", in *1st Int. Symp. on Object-Oriented Real-Time Distributed Computing (ISORC'98)*, (Kyoto, Japan), pp.174-83, IEEE Computer Society Press, 1998.

[Bondavalli *et al.* 2000] A. Bondavalli, S. Chiaradonna, F. D. Giandomenico and F. Grandoni, "Threshold-based Mechanisms to Discriminate Transient from Intermittent Faults", *IEEE Transactions on Computers*, 49 (3), pp.230-45, 2000.

[Bouali *et al.* 1994] A. Bouali, S. Gnesi and S. Larosa, "The Integration Project for the JACK Environment", *Bulletin of the European Association for Theoretical Computer Science*, October (54), pp.207-23, 1994.

[Brownbridge *et al.* 1982] D. R. Brownbridge, L. F. Marshall and B. Randell, "The Newcastle Connection, or - UNIXes of the World Unite!", *Software Practice and Experience*, 12 (12), pp.1147-62, 1982.

[Burns & Wellings 1993] A. Burns and A. Wellings, *HRT-HOOD: A Structured Design Method for Hard Real-Time Ada Systems - v2.0 Reference Manual*, 1993.

[Burns & Wellings 1994] A. Burns and A. Wellings, "HRT-HOOD: A Design Method for Hard Real-Time Ada", *Real-Time Systems*, 6 (1), pp.73-114, 1994.

[Burns & Wellings 1995a] A. Burns and A. J. Wellings, "Safety Kernels: Specification and Implementation", *High Integrity Systems*, 3 (1), pp.287-300, 1995.

[Burns & Wellings 1995b] A. Burns and A. Wellings, *Hard Real-Time HOOD: A Structured Design Method for Hard Real-Time Ada Systems,* Real-Time Safety Critical Systems, 3, 313p., Elsevier, 1995.

[Burns & Wellings 1996] A. Burns and A. J. Wellings, *Real-Time Systems and their Programming Languages,* Addison Wesley, 1996.

[Burns *et al.* 1995] A. Burns, N. Hayes and M. F. Richardson, "Generating Feasible Cyclic Schedules", *Control Engineering Practice*, 3 (2), pp.151-62, 1995.

[Cachin *et al.* 2000] C. Cachin, J. Camenisch, M. Dacier, Y. Deswarte, J. Dobson, D. Horne, K. Kursawe, J.-C. Laprie, J.-C. Lebraud, D. Long, T. McCutcheon, J. Müller, F. Petzold, B. Pfitzmann, D. Powell, B. Randell, M. Schunter, V. Shoup, P. Veríssimo, G. Trouessin, R. Stroud, M. Waidner and I. Welch, *Reference Model and Use Cases*, Report N°00280, LAAS-CNRS, August 2000 (Project IST-1999-11583: MAFTIA).

[Canver *et al.* 1997] E. Canver, D. Schwier and F. W. von Henke, *Formal Specification and Validation of the Clock Synchronization Algorithm,* ESPRIT Project 20716 GUARDS Report N°D3A5.A0.0801.C, University of Ulm, Germany, March 1997.

[Carreira *et al.* 1998] J. Carreira, H. Madeira and J. G. Silva, "Xception: A Technique for the Experimental Evaluation of Dependability in Modern Computers", *IEEE Transactions on Software Engineering*, 24 (2), pp.125-36, 1998.

[Chiba 1995] S. Chiba, "A Metaobject Protocol for C++", in *ACM Conf. on Object-Oriented Programming, Systems, Languages and Applications,* (Austin, TX, USA), pp.285-99, 1995.

[Clark & Wilson 1987] D. D. Clark and D. R. Wilson, "A Comparison of Commercial and Military Computer Security Policies", in *Symp. on Security and Privacy,* (Oakland, CA, USA), pp.184-94, IEEE Computer Society Press, 1987.

[Clarke *et al.* 1986] E. M. Clarke, E. A. Emerson and E. Sistla, "Automatic Verification of Finite State Concurrent Systems using Temporal Logic Specifications: A Practical Approach", *ACM Transactions on Programming Languages and Systems*, 8 (2), pp.244-63, 1986.

[Cristian 1989] F. Cristian, "Probabilistic Clock Synchronization", *Distributed Computing,* 3 (3), pp.146-58, 1989.

[De Nicola & Vaandrager 1990] R. De Nicola and F. W. Vaandrager, "Actions versus State Based Logics for Transition Systems", in *LITP Spring School on Semantics of Concurrency,* (La Roche Posay, France) Lecture Notes in Computer Science, 469, pp.407-19, Springer-Verlag, Berlin, 1990.

[Dolev *et al.* 1995] D. Dolev, J. Y. Halpern, B. Simons and R. Strong, "Dynamic Fault-Tolerant Clock Synchronization", *Journal of the ACM,* 42 (1), pp.143-85, 1995.

[Dutuit *et al.* 1997] Y. Dutuit, E. Châtelet, J.-P. Signoret and P. Thomas, "Dependability Modelling and Evaluation by Using Stochastic Petri Nets: Application to Two Test Cases", *Reliability Engineering and System Safety,* 55, pp.117-24, 1997.

[Emerson & Halpern 1986] E. A. Emerson and J. Y. Halpern, "'Sometimes' and 'Not Never' Revisited: On Branching versus Linear Time Temporal Logic", *Journal of the ACM,* 33 (1), pp.151-78, 1986.

[EN 50170] *General Purpose Field Communication System,* Comité Européen de Normalisation Électrotechnique (CENELEC), June 1997.

[ESA 1991] ESA, *Hood Reference Manual Issue 3.1,* N°HRM/91-07/V3.1, European Space Agency, 1991.

[Fabre & Pérennou 1998] J.-C. Fabre and T. Pérennou, "A Metaobject Architecture for Fault-Tolerant Distributed Systems: The FRIENDS Approach", *IEEE Transactions on Computers,* 47 (1), pp.78-95, 1998.

[Fantechi *et al.* 1998] A. Fantechi, S. Gnesi, F. Mazzanti, R. Pugliese and E. Tronci, "A Symbolic Model Checker for ACTL", in *FM-Trends'98: Int. Workshop on Current Trends in Applied Formal Methods,* (Boppard, Germany), Lecture Notes on Computer Science, 1641, pp.228-42, Springer-Verlag, 1998.

[Fantechi *et al.* 1999] A. Fantechi, S. Gnesi and L. Semini, "Formal Description and Validation for an Integrity Policy Supporing Multiple Levels of Criticality", in *Dependable Computing for Critical Applications (DCCA-7),* (C. B. Weinstock and J. Rushby, Eds.), Dependable Computing and Fault-Tolerant Systems, 12, pp.129-46, IEEE Computer Society Press, 1999.

[Gallmeister 1995] B. O. Gallmeister, *POSIX.4,* O'Reilly&Associates, 1995.

[Garey & Johnson 1979] M. R. Garey and D. S. Johnson, *Computers and Intractability,* Freeman, New York, 1979.

[Gong *et al.* 1998] L. Gong, P. Lincoln and J. Rushby, "Byzantine Agreement with Authentication: Observations and Applications in Tolerating Hybrid and Link Faults", in *Dependable Computing for Critical Applications (DCCA-5),* (R. K. Iyer, M. Morganti, W. K. Fuchs and V. Gligor, Eds.), Dependable Computing and Fault-Tolerant Systems, 10, pp.139-57, IEEE Computer Society Press, 1998.

[Grandoni *et al.* 1998] F. Grandoni, S. Chiaradonna and A. Bondavalli, "A New Heuristic to Discriminate between Transient and Intermittent Faults", in *3rd IEEE High Assurance System Engineering Symposium (HASE),* (Bethesda, MD, USA), pp.224-31, IEEE Computer Society Press, 1998.

[Gray 1986] J. Gray, "Why do Computers Stop and What can be done about it?", in *5th Symp. on Reliability in Distributed Software and Database Systems,* (Los Angeles, CA, USA), pp.3-12, IEEE Computer Society Press, 1986.

[Harper & Lala 1990] R. E. Harper and J. H. Lala, "Fault-Tolerant Parallel Processor", *Journal of Guidance, Control and Dynamics,* 14 (3), pp.554-63, 1990.

[Hoyme & Driscoll 1992] K. Hoyme and K. Driscoll, "SAFEbus®", in *11th Digital Avionics Systems Conference,* (Seattle, WA, USA), pp.68-73, AIAA/IEEE, 1992.

[Hsueh *et al.* 1997] M.-C. Hsueh, T. K. Tsai and R. K. Iyer, "Fault Injection Techniques and Tools", *IEEE Computer,* 40 (4), pp.75-82, 1997.

[Hugue & Stotts 1991] M. C. E. Hugue and P. D. Stotts, "Guaranteed Task Deadlines for Fault-Tolerant Workloads with Conditional Branches", *Real-Time Systems,* 3 (3), pp.275-305, 1991.

[IEC 61226] *Nuclear Power Plants - Instrumentation and Control Systems Important for Safety - Classification*, International Electrotechnical Commission, May 1993.

[IEC 61508] *Functional Safety of Electrical/Electronic/Programmable Electronic Safety-Related Systems*, International Electrotechnical Commission, December 1998.

[IEEE 1003] *Portable Operating System Interface*, Institute of Electrical and Electronic Engineers, 1990.

[IEEE 1003.1] *Portable Operating System Interface: Part 1: System Application Program Interface (API) [C language]*, Institute of Electrical and Electronic Engineers, 1990.

[IEEE 1003.1b] *Portable Operating System Interface: Amendment 1 — Realtime Extension [C language]*, Institute of Electrical and Electronic Engineers, 1993.

[IEEE 1003.1c] *Portable Operating System Interface: Amendment 2 — Threads Extension [C language]*, Institute of Electrical and Electronic Engineers, 1995.

[Intecs-Sistemi 1996] *HRT-HoodNICE: a Hard Real-Time Software Design Support Tool*, Intecs Sistemi, Pisa, Italy, ESTEC Contract 11234/NL/FM(SC), Final Report, 1996 (see also http://www.pisa.intecs.it/products/HRT-HoodNICE/).

[Intecs-Sistemi 1997] *HoodNICE v2.3.4 Reference Manual and User Guide*, Intecs Sistemi, Pisa, Italy, 1997.

[Iyer *et al.* 1990] R. K. Iyer, L. T. Young and P. V. K. Iyer, "Automatic Recognition of Intermittent Failures: An Experimental Study of Field Data", *IEEE Transactions on Computers*, C-39 (4), pp.525-37, 1990.

[Jajodia & Kogan 1990] S. Jajodia and B. Kogan, "Integrating an Object-Oriented Data Model with Multilevel Security", in *Symp. on Security and Privacy*, (Oakland, CA, USA), pp.76-85, IEEE Computer Society Press, 1990.

[Jenn 1998a] E. Jenn, *Modelling for Evaluation*, ESPRIT Project 20716 GUARDS Report N°I3A3.TN.0056.A, Technicatome, Aix-en-Provence, France, January 1998.

[Jenn 1998b] E. Jenn, *Preliminary Design of the Modelling Environment*, ESPRIT Project 20716 GUARDS Report N°I3A3.TN.0044.A, Technicatome, Aix-en-Provence, France, June 1998.

[Jenn 1999] E. Jenn, *Nuclear Instance Implementation Report*, ESPRIT Project 20716 GUARDS Report N°A2A1.TN.0113A, Technicatome, Aix-en-Provence, France, March 1999.

[Joseph & Pandya 1986] M. Joseph and P. Pandya, "Finding Response Times in a Real-Time System", *BCS Computer Journal*, 29 (5), pp.390-5, 1986.

[Kanoun & Borrel 2000] K. Kanoun and M. Borrel, "Fault-Tolerant System Dependability — Explicit Modeling of Hardware and Software Component Interactions", *IEEE Transactions on Reliability*, 49 (3) 2000.

[Kanoun *et al.* 1999] K. Kanoun, M. Borrel, T. Moreteveille and A. Peytavin, "Modeling the Dependability of CAUTRA, a Subset of the French Air Traffic Control System", *IEEE Transactions on Computers*, 48 (5), pp.528-35, 1999.

[Kiczales *et al.* 1991] G. Kiczales, J. des Rivières and D. G. Bobrow, *The Art of the Metaobject Protocol*, MIT Press, 1991.

[Kieckhafer *et al.* 1988] R. M. Kieckhafer, C. J. Walter, A. M. Finn and P. M. Thambidurai, "The MAFT Architecture for Distributed Fault Tolerance", *IEEE Transactions on Computers*, 37 (4), pp.398-405, 1988.

[Kopetz 1997] H. Kopetz, "Component-Based Design of Large Distributed Real-Time Systems", in *IFAC Workshop on Distributed Computer Control Systems (DCCS 97)*, (Seoul, Korea), 1997.

[Krishnamurthy *et al.* 1998] N. Krishnamurthy, V. Jhaveri and J. A. Abraham, "A Design Methodology for Software Fault Injection in Embedded Systems", in *Proc. IFIP Int. Workshop on Dependable Computing and Its Applications (DCIA-98)*, (Y. Chen, Ed.), (Johannesburg, South Africa), pp.237-48, 1998.

[Lala & Alger 1988] J. H. Lala and L. S. Alger, "Hardware and Software Fault Tolerance: a Unified Architectural Approach", in *18th IEEE Int. Symp. on Fault Tolerant Computing (FTCS-18),* (Tokyo, Japan), pp.240-5, IEEE Computer Society Press, 1988.

[Lala & Harper 1994] J. H. Lala and R. E. Harper, "Architectural Principles for Safety-Critical Real-Time Applications", *Proceedings of the IEEE,* 82 (1), pp.25-40, 1994.

[Lamport & Melliar-Smith 1985] L. Lamport and P. M. Melliar-Smith, "Synchronizing Clocks in the Presence of Faults", *Journal of the ACM,* 32 (1), pp.52-78, 1985.

[Lamport *et al.* 1982] L. Lamport, R. Shostak and M. Pease, "The Byzantine Generals Problem", *ACM Transactions on Programming Languages and Systems,* 4 (3), pp.382-401, 1982.

[Laprie 1992] J.-C. Laprie (Ed.), *Dependability: Basic Concepts and Terminology,* Dependable Computing and Fault-Tolerance, 5, 265p., Springer-Verlag, Vienna, Austria, 1992.

[Laprie 1995] J.-C. Laprie, "Dependability of Computer Systems: Concepts, Limits, Improvements", in *6th Int. Symp. on Software Reliability Engineering (ISSRE'95),* (Toulouse, France), pp.2-11, IEEE Computer Society Press, 1995.

[Laprie *et al.* 1995] J.-C. Laprie, J. Arlat, J.-P. Blanquart, A. Costes, Y. Crouzet, Y. Deswarte, J.-C. Fabre, H. Guillermain, M. Kaâniche, K. Kanoun, C. Mazet, D. Powell, C. Rabéjac and P. Thévenod-Fosse, *Dependability Guidebook,* 324p., Cépaduès-Editions, Toulouse, 1995 (in French).

[Lee & Shin 1990] S. Lee and K. G. Shin, "Optimal Multiple Syndrome Probabilistic Diagnosis", in *20th IEEE Int. Symp. on Fault-Tolerant Computing Systems (FTCS-20),* (Newcastle upon Tyne, UK), pp.324-31, IEEE Computer Society Press, 1990.

[Leung & Whitehead 1982] J. Leung and J. Whitehead, "On the Complexity of Fixed-Priority Scheduling of Periodic, Real-Time Tasks", *Performance Evaluation,* 2 (4), pp.237-50, 1982.

[Lin & Siewiorek 1990] T.-T. Y. Lin and D. P. Siewiorek, "Error Log Analysis: Statistical Modeling and Heuristic Trend Analysis", *IEEE Transactions on Reliability,* 39 (4), pp.419-32, 1990.

[Lincoln & Rushby 1993] P. Lincoln and J. Rushby, "A Formally Verified Algorithm for Interactive Consistency Under a Hybrid Fault Model", in *23rd IEEE Int. Symp. on Fault-Tolerant Computing (FTCS-23),* (Toulouse, France), pp.402-11, IEEE Computer Society Press, 1993.

[Liu & Layland 1973] C. L. Liu and J. W. Layland, "Scheduling Algorithms for Multiprogramming in a Hard Real-Time Environment", *Journal of the ACM,* 20 (1), pp.46-61, 1973.

[Lundelius-Welch & Lynch 1988] J. Lundelius-Welch and N. Lynch, "A New Fault-Tolerant Algorithm for Clock Synchronization", *Information and Computation,* 77 (1), pp.1-16, 1988.

[Melliar-Smith & Schwartz 1982] P. M. Melliar-Smith and R. L. Schwartz, "Formal Specification and Mechanical Verification of SIFT: A Fault-Tolerant Flight Control System", *IEEE Transactions on Computers,* C-31 (7), pp.616-30, 1982.

[Mongardi 1993] G. Mongardi, "Dependable Computing for Railway Control Systems", in *Dependable Computing for Critical Applications (DCCA-3),* (C. E. Landwehr, B. Randell and L. Simoncini, Eds.), Dependable Computing and Fault-Tolerant Systems, 8, pp.255-77, Springer-Verlag, Vienna, Austria, 1993.

[Mura & Bondavalli 1997] I. Mura and A. Bondavalli, *Hierarchical Modelling and Evaluation of Phased-Mission Systems,* ESPRIT Project 20716 GUARDS Report N°I3A.AO.6007.A, PDCC, Pisa, Italy, November 1997.

[Mura & Bondavalli 1999] I. Mura and A. Bondavalli, "Hierarchical Modelling and Evaluation of Phased-Mission Systems", *IEEE Transactions on Reliability,* 48 (4), pp.360-8, 1999.

[NCSC 1987] *Trusted Network Interpretation of the Trusted Computer System Evaluation Criteria,* National Computer Security Center, 1987.

[OMG 1995] *The Common Object Request Broker: Architecture and Specification,* Object Management Group, 1995 (revision 2.0).

[Oswald & Attermeyer 1999] D. Oswald and B. Attermeyer, *Software Implemented Fault Injection for GUARDS: FITS Software User Manual*, ESPRIT Project 20716 GUARDS Report N°I3A2.AO.4009.A, Siemens AG Österreich, Vienna, Austria, June 1999.

[Oswald *et al.* 1999] D. Oswald, C. Nagy, M. Jantsch and B. Attermeyer, *Software Implemented Fault Injection for GUARDS: FITS Detailed Design*, ESPRIT Project 20716 GUARDS Report N°I3A2.AO.4008.C, Siemens AG Österreich, Vienna, Austria, June 1999.

[Owre *et al.* 1996] S. Owre, S. Rajan, J. M. Rushby, N. Shankar and M. K. Srivas, "PVS: Combining Specification, Proof Checking, and Model Checking", in *Computer-Aided Verification (Proc. CAV'96, New Brunswick, NJ, USA, July/August 1996)*, (R. Alur and T. A. Henzinger, Eds.), Lecture Notes in Computer Science, 1102, pp.411-4, Springer-Verlag, New-York, USA, 1996.

[Paganone & Coppola 1997] A. Paganone and P. Coppola, *Specification and Preliminary Design of the Architectural Development Environment*, ESPRIT Project 20716 GUARDS Report N°D2A1.A0.3002.C, Intecs Sistemi, Pisa, Italy, April 1997.

[Pease *et al.* 1980] M. Pease, R. Shostak and L. Lamport, "Reaching Agreement in the Presence of Faults", *Journal of the ACM*, 27 (2), pp.228-34, 1980.

[Poledna 1998] S. Poledna, "Deterministic Operation in of Dissimilar Replicated Tasks Sets in Fault-Tolerant Distributed Real-Time Systems", in *Dependable Computing for Critical Applications (DCCA-6)*, (M. Dal Cin, C. Meadows and W. H. Sanders, Eds.), pp.103-19, IEEE Computer Society Press, 1998.

[Powell 1994] D. Powell, "Distributed Fault-Tolerance — Lessons from Delta-4", *IEEE Micro*, 14 (1), pp.36-47, 1994.

[Powell 1997] D. Powell, *Preliminary Definition of the GUARDS Architecture*, ESPRIT Project 20716 GUARDS Report N°D1A1.A0.5000.D, LAAS-CNRS, Toulouse, France, January 1997 (LAAS Research Report N°98136).

[Powell *et al.* 1998a] D. Powell, C. Rabéjac and A. Bondavalli, *Alpha-count Mechanism and Inter-Channel Diagnosis*, ESPRIT Project 20716 GUARDS Report, N°I1-SA1.TN.5009.E, LAAS-CNRS, Toulouse, France, February 1998.

[Powell *et al.* 1998b] D. Powell, J. Arlat and K. Kanoun, *Generic Architecture Instantiation Guidelines*, ESPRIT Project 20716 GUARDS Report N°I1SA1.TN.5008.C, LAAS-CNRS, Toulouse, France, May 1998 (LAAS Research Report N°98136).

[Rabéjac 1997] C. Rabéjac, *Inter-Channel Fault Treatment Mechanism*, ESPRIT Project 20716 GUARDS Report N°D1.A3.AO.2014.B, Matra Marconi Space, Toulouse, France, March 1997.

[Rajkumar *et al.* 1994] R. Rajkumar, L. Sha, J. P. Lehoczky and K. Ramamritham, "An Optimal Priority Inheritance Policy for Synchronization in Real-Time Systems", in *Advances in Real-Time Systems*, (S. H. Son, Ed.), pp.249-71, Prentice-Hall, 1994.

[Ramanathan *et al.* 1990] P. Ramanathan, K. G. Shin and R. W. Butler, "Fault-Tolerant Clock Synchronization in Distributed Systems", *Computer*, pp.33-42, 1990.

[Rivest 1992] R. L. Rivest, *The MD5 Message-Digest Algorithm*, RFC N°1321, MIT Laboratory for Computer Science and RSA Data Security, Inc., April 1992.

[Rivest *et al.* 1978] R. L. Rivest, A. Shamir and L. Adleman, "A Method for Obtaining Digital Signatures and Public-Key Cryptosystems", *Communications of the ACM*, 21 (2), pp.120-6, 1978.

[Robinson 1992] P. Robinson, *Hierarchical Object-Oriented Design*, Prentice-Hall, 1992.

[Rozier *et al.* 1990] M. Rozier, V. Abrossimov, F. Armand, I. Boule, M. Gien, M. Guillemont, F. Herrmann, C. Kaiser, S. Langlois, P. Léonard and W. Neuhauser, *Overview of the CHORUS® Distributed Operating Systems*, Report N°CS/TR-90-25, Chorus Systèmes, April 1990.

[Sanders & Meyer 1991] W. H. Sanders and J. F. Meyer, "A Unified Approach for Specifying Measures of Performance, Dependability and Performability", in *Dependable Computing for Critical Applications (DCCA-1)*, (A. Avizienis and J.-C. Laprie, Eds.), Dependable Computing and Fault-Tolerant Systems, 4, pp.215-37, Springer-Verlag, Vienna, Austria, 1991.

[Sanders *et al.* 1995] W. H. Sanders, W. D. Obal, M. A. Qureshi and F. K. Widjanarko, "The UltraSAN Modelling Environment", *Performance Evaluation Journal*, 24, pp.89-115, 1995.

[Schwier & von Henke 1997] D. Schwier and F. von Henke, "Mechanical Verification of Clock Synchronization Algorithms", in *Design for Validation*, ESPRIT Long Term Research Project 20072: DeVa - 2nd Year Report, pp.287-303, LAAS-CNRS, Toulouse, France, 1997.

[Semini 1998] L. Semini, *Formal Specification and Verification for an Integrity Policy Supporting Multiple Levels of Criticality*, ESPRIT Project 20716 GUARDS Report N°I3A5.AO.6012.A, PDCC, Pisa, Italy, April 1998.

[Sha *et al.* 1990] L. Sha, R. Rajkumar and J. P. Lehoczky, "Priority Inheritance Protocols: An Approach to Real-Time Synchronization", *IEEE Transactions on Computers*, 39 (9), pp.1175-85, 1990.

[Simpson 1990] H. Simpson, "Four-Slot Fully Asynchronous Communication Mechanism", *IEE Proceedings*, 137, Part E (1), pp.17-30, 1990.

[Sims 1996] T. Sims, "Real Time Recovery of Fault Tolerant Processing Elements", in *AIAA/IEEE 15th Digital Avionics Systems Conference*, (Atlanta, GA, USA), pp.485-90, IEEE, 1996.

[Spainhower *et al.* 1992] L. Spainhower, J. Isenberg, R. Chillarege and J. Berding, "Design for Fault-Tolerance in System ES/9000 Model 900", in *22nd IEEE Int. Symp. on Fault-Tolerant Computing Systems (FTCS-22)*, (Boston, MA, USA), pp.38-47, IEEE Computer Society Press, 1992.

[Srikanth & Toueg 1987] T. K. Srikanth and S. Toueg, "Optimal Clock Synchronization", *Journal of the ACM*, 34 (3), pp.626-45, 1987.

[Sun Microsystems 1989] Sun Microsystems, *NFS: Network File System Protocol Specification*, N°RFC-1094, March 1989.

[Tendolkar & Swann 1982] N. N. Tendolkar and R. L. Swann, "Automated Diagnostic Methodology for the IBM 3081 Processor Complex", *IBM Journal of Research and Development*, 26 (1), pp.78-88, 1982.

[Thambidurai & Park 1988] P. Thambidurai and Y.-K. Park, "Interactive Consistency with Multiple Failure Modes", in *7th Symp. on Reliable Distributed Systems (SRDS-7)*, (Columbus, OH, USA), pp.93-100, IEEE Computer Society Press, 1988.

[Tindell 1993] K. Tindell, *Fixed Priority Scheduling of Hard Real-Time Systems*, D. Phil. Thesis, Dept. of Computer Science, University of York, UK, 1993.

[Totel 1998] E. Totel, *Multilevel Integrity Policy for Run-Time Protection of Critical Tasks*, Doctorate Thesis, Institut National Polytechnique de Toulouse, 1998 (in French).

[Totel *et al.* 1998] E. Totel, J.-P. Blanquart, Y. Deswarte and D. Powell, "Supporting Multiple Levels of Criticality", in *28th IEEE Int. Symp. on Fault-Tolerant Computing (FTCS-28)*, (Munich, Germany), pp.70-9, IEEE Computer Society Press, 1998.

[Tsao & Siewiorek 1983] M. M. Tsao and D. P. Siewiorek, "Trend Analysis on System Error Files", in *13th IEEE Int. Symp. on Fault-Tolerant Computing (FTCS-13)*, (Milan, Italy), pp.116-9, IEEE Computer Society Press, 1983.

[Wellings *et al.* 1998] A. Wellings, L. Beus-Dukic and D. Powell, "Real-Time Scheduling in a Generic Fault-Tolerant Architecture", in *19th Real-Time Systems Symp. (RTSS-19)*, (Madrid, Spain), pp.390-8, IEEE Computer Society Press, 1998.